Cloud Native
Software Security Handbook

Unleash the power of cloud native tools for robust security
in modern applications

Mihir Shah

BIRMINGHAM—MUMBAI

Cloud Native Software Security Handbook

Group Product Manager: Preet Ahuja
Publishing Product Manager: Suwarna Rajput
Book Project Manager: Ashwin Dinesh Kharwa
Content Development Editor: Sujata Tripathi
Technical Editor: Arjun Varma
Copy Editor: Safis Editing
Proofreader: Safis Editing
Indexer: Subalakshmi Govindhan
Production Designer: Ponraj Dhandapani
DevRel Marketing Coordinator: Rohan Dobhal

First published: August 2023

Production reference: 1270723

Published by Packt Publishing Ltd.
Grosvenor House
11 St Paul's Square
Birmingham
B3 1RB

ISBN 9781837636983

www.packtpub.com

To all who dare to dream: you can, if you believe you can.

– Mihir Shah

Contributors

About the author

Mihir Shah is a recognized industry expert in the cybersecurity domain. He has been a speaker at premier academic institutes, such as Stanford University and IIT Bombay. He was an industry mentor for Stanford University's Advanced Cybersecurity program, where he worked with multiple start-up stakeholders as a mentor. He is an active researcher at MIT and has published several papers over the past few years. He was invited to be a judge and industry expert for the coveted Globee Business Excellence awards for the Cybersecurity category. He has delivered over 30 talks over the past five years at seven conferences around the globe. He has several years of industry experience working as a software security engineer, leading the product security division for cloud engineering teams.

About the reviewers

Kuldeep Singh is an experienced data center migration architect and leader. He has expertise in infrastructure project management and has delivered successful data center projects for telecom giants such as AT&T (US) and Vodafone (Europe). He also holds various certifications for cloud solutions and project management. In his free time, he likes to keep fit by lifting weights at the gym.

I am grateful to my wife, my kid, Mozo, and my family for their constant support and encouragement throughout this project. I also appreciate the publisher for their professionalism and guidance in the publishing process. I hope this book will be useful and enjoyable to readers.

Safeer CM has been working in site reliability, DevOps, and platform engineering for the past 17 years. A site reliability engineer by trade, Safeer has managed large-scale production infrastructures at internet giants such as Yahoo and LinkedIn and is currently working at Flipkart as a senior staff site reliability engineer. He has worked with budding and established start-ups as a cloud architect and DevOps/SRE consultant.

Safeer is the author of the book *Architecting Cloud-Native Serverless Solutions*, as well as several blogs. He has been a speaker at and organizer of multiple meetups. He is currently an ambassador for the Continuous Delivery Foundation, where he helps the organization with community adoption and governance.

All my contributions were made possible by the infinite support of my family. I would also like to acknowledge the wisdom and support of all my mentors, peers, and the technology community around the world.

Aditya Krishnakumar has worked in the field of DevOps for the past 5+ years, with 2 years working with organizations on security-related requirements such as SOC 2. He is currently working as a senior infrastructure engineer at Sysdig, the makers of the Falco runtime security platform. He previously worked for **Boston Consulting Group** (**BCG**), where he was responsible for the SOC 2 requirements of a big data analytics product hosted on Kubernetes. He provides contributions to the DevOps community via his blog and is a member of AWS Community Builders, a global community of AWS and cloud enthusiasts.

Mayur Nagekar is an accomplished professional with 15+ years of experience in DevOps, automation, and platform engineering, and is passionate about cloud-native technologies. As a site reliability engineer and cloud architect, he has extensive expertise in Infrastructure as Code, designing distributed systems, and implementing cloud-native applications. Mayur holds a bachelor's degree in computer science from the University of Pune, and his experience spans from start-ups to multinational companies. His proficiency in Kubernetes, containerization, and security drives innovation and efficiency.

I am incredibly grateful to my loving family and supportive friends who have been with me every step of the way. Your unwavering encouragement, understanding, and belief in me have made reviewing this book possible. Thank you for your love, patience, and unwavering support. I am truly blessed to have you all in my life.

Table of Contents

3

Cloud Native Application Security   69

Part 2: Implementing Security in Cloud Native Environments

4

Building an AppSec Culture   107

5

Threat Modeling for Cloud Native   137

6

Securing the Infrastructure   181

7

Cloud Security Operations   217

8

DevSecOps Practices for Cloud Native   253

Part 3: Legal, Compliance, and Vendor Management

9

Legal and Compliance   285

10

Cloud Native Vendor Management and Security Certifications   307

Preface

Writing the *Cloud Native Software Security Handbook* has been an exciting and fulfilling journey for me. As an author, I am passionate about helping you navigate the complex world of cloud-native security, equipping you with the knowledge and skills necessary to secure infrastructure and develop secure software in this rapidly evolving landscape.

Throughout my experience in the field, I have witnessed the transformative power of cloud-native technologies and their potential to revolutionize the way we build and deploy software. However, I have also come to realize the critical importance of robust security practices in this domain. It is this realization that motivated me to write this book – to bridge the gap between the power of cloud-native platforms and the need for comprehensive security measures.

As I delved into the creation of this handbook, I considered the needs of those among you who are eager to explore the cloud-native space and embrace its potential, while ensuring the utmost security. I embarked on a deep dive into widely used platforms such as Kubernetes, Calico, Prometheus, Kibana, Grafana, Clair, and Anchor, and many others – equipping you with the tools and knowledge necessary to navigate these technologies with confidence.

Beyond the technical aspects, I wanted this book to be a guide that goes beyond the surface and addresses the broader organizational and cultural aspects of cloud-native security. In the latter part of this book, we will explore the concept of **Application Security** (**AppSec**) programs and discuss how to foster a secure coding culture within your organization. We will also dive into threat modeling for cloud-native environments, empowering you to proactively identify and mitigate potential security risks.

Throughout this journey, I have strived to present practical insights and real-world examples that will resonate with those of you from diverse backgrounds. I believe that by sharing both my own experiences and those of others in the field, we can cultivate a sense of camaraderie and mutual growth as we navigate the intricacies of cloud-native security together.

My hope is that by the end of this book, you will not only possess a comprehensive understanding of cloud-native security but also feel confident in your ability to create secure code and design resilient systems. I invite you to immerse yourself in this exploration, embrace the challenges, and seize the opportunities that await you in the realm of cloud-native software security.

Who this book is for

This book is intended for developers, security professionals, and DevOps teams who are involved in designing, developing, and deploying cloud-native applications. It is particularly beneficial for those with a technical background who wish to gain a deeper understanding of cloud-native security and learn about the latest tools and technologies, to secure cloud-native infrastructure and runtime environments. Prior experience with cloud vendors and their managed services would be advantageous.

What this book covers

Chapter 1, Foundations of Cloud Native, serves as a comprehensive introduction to cloud-native technologies, exploring the tools and platforms offered by the CNCF. It provides a clear understanding of these platforms, their use cases and applications, and how to deploy them in real time. It is designed to help those of you who are familiar with public cloud vendors and their offerings but seek to understand how they integrate with vendor-agnostic cloud-native technologies.

Chapter 2, Cloud Native Systems Security Management, provides a comprehensive understanding of the various tools and techniques that can be used to secure cloud-native systems, and how they can be integrated to provide a secure and compliant cloud-native environment. By the end of this chapter, you will be able to implement secure configuration management, secure image management, secure runtime management, secure network management, and Kubernetes admission controllers in their cloud-native systems.

Chapter 3, Cloud Native Application Security, provides an in-depth understanding of the security considerations involved in cloud-native application development. As the shift toward cloud-based application development continues to grow, it is crucial for software engineers, architects, and security professionals to understand the security implications and best practices to build secure cloud-native applications.

Chapter 4, Building an AppSec Culture, covers the key components of building an AppSec program that is both effective and efficient. It emphasizes the importance of understanding your organization's security needs and goals and explores the key elements of an effective AppSec program, including risk assessment, security testing, and security training.

Chapter 5, Threat Modeling for Cloud Native, provides a comprehensive understanding of how to perform threat modeling for cloud-native environments, and how to use the information gathered to make informed decisions about security risks. It brings together all the concepts covered so far and applies them to the process of threat modeling.

Chapter 6, Securing the Infrastructure, explores various tools and strategies to secure your cloud-native infrastructure, from configuration to network security. It provides hands-on experience in implementing various security measures for Kubernetes, service mesh, and container security.

Chapter 7, Cloud Security Operations, offers practical insights and tools to establish and maintain a robust cloud security operations process. It explores innovative techniques to collect and analyze data points, including centralized logging, cloud-native observability tools, and monitoring with Prometheus and Grafana.

Chapter 8, DevSecOps Practices for Cloud Native, delves into the various aspects of DevSecOps, focusing on **Infrastructure as Code (IaC)**, policy as code, and **Continuous Integration/Continuous Deployment (CI/CD)** platforms. This chapter will teach you in detail about automating most of the processes you learned in the previous chapters. By the end of this chapter, you will have a comprehensive understanding of these concepts and the open source tools that aid in implementing DevSecOps practices.

Chapter 9, Legal and Compliance, aims to bridge the gap between the technical skills and the legal and compliance aspects in the world of cloud-native software security. This chapter provides you with a comprehensive understanding of the laws, regulations, and standards that govern your work. By the end of this chapter, you will not only gain knowledge about the key U.S. privacy and security laws but also learn how to analyze these laws from a security engineer's perspective.

Chapter 10, Cloud Native Vendor Management and Security Certifications, dives deep into the world of cloud vendor management and security certifications, revealing practical tools and strategies to build strong vendor relationships that underpin secure cloud operations. By the end of this chapter, you will understand the various risks associated with cloud vendors and how to assess a vendor's security posture effectively.

To get the most out of this book

Before starting with this book, it is expected that you have a preliminary understanding of cloud-native technologies such as Kubernetes and Terraform. This book was written to explain security solutions possible using the following cloud-native tools, and so it is expected that you should adopt a security mindset when learning about the tools or using them. This book has a lot of examples and references for you to follow and implement; it is expected that you don't use the code, as provided, verbatim, as each environment is different. Instead, approach each chapter carefully, and apply your learnings in your own environment. I hope that you spend more time learning about the tool itself, as that provides a holistic understanding of what this book aims to achieve – cloud-native security.

Software/hardware covered in the book	Operating system requirements
Kubernetes v 1.27	macOS or Linux
Helm v3.12.0	macOS or Linux
Open Policy Agent v 0.52.0	macOS or Linux
Harbor v 2.7.0	macOS or Linux
Clair v 4.6.0	macOS or Linux

K9s v 0.27.2	macOS or Linux
Vault v 1.13.2	macOS or Linux
OWASP ASVS v 4.0	macOS or Linux
Calico v 3.25	macOS or Linux
Falco	macOS or Linux
OPA – Gatekeeper v 3.10	macOS or Linux
Elasticsearch v 7.13.0	macOS or Linux
Fluentd v 1.15.1	macOS or Linux
Kibana v 8.7.0	macOS or Linux
Prometheus v 2.44.0	macOS or Linux
Terraform v 1.4.6	macOS or Linux
Checkov v 2.3.245	macOS or Linux

For certain tools, where the installation guide is a little complex, steps and tutorials are included within each chapter; however, you are strongly advised to follow the official documentation to install the tools as listed in the preceding table before trying the hands-on tutorials.

If you are using the digital version of this book, we advise you to type the code yourself or access the code from the book's GitHub repository (a link is available in the next section). Doing so will help you avoid any potential errors related to the copying and pasting of code.

Download the example code files

You can download the example code files for this book from GitHub at `https://github.com/PacktPublishing/Cloud-Native-Software-Security-Handbook`. If there's an update to the code, it will be updated in the GitHub repository.

We also have other code bundles from our rich catalog of books and videos available at `https://github.com/PacktPublishing/`. Check them out!

Conventions used

There are a number of text conventions used throughout this book.

`Code in text`: Indicates code words in text, database table names, folder names, filenames, file extensions, pathnames, dummy URLs, user input, and Twitter handles. Here is an example: "You should receive an error message indicating that the namespace must have the `environment` label. Update the `test-namespace.yaml` file to include the required label, and the namespace creation should be allowed."

A block of code is set as follows:

```
kind: NetworkPolicy
apiVersion: networking.k8s.io/v1
metadata:
  name: frontend-to-backend
spec:
  podSelector:
    matchLabels:
      app: backend
  policyTypes:
  - Ingress
  ingress:
  - from:
    - podSelector:
        matchLabels:
          app: frontend
    ports:
    - protocol: TCP
      port: 80
---
kind: NetworkPolicy
apiVersion: networking.k8s.io/v1
metadata:
  name: backend-to-database
spec:
  podSelector:
    matchLabels:
      app: database
  policyTypes:
  - Ingress
  ingress:
  - from:
    - podSelector:
        matchLabels:
          app: backend
    ports:
    - protocol: TCP
      port: 3306
```

Any command-line input or output is written as follows:

```
$ kubectl apply -f networkPolicy.yaml
```

Bold: Indicates a new term, an important word, or words that you see onscreen. For instance, words in menus or dialog boxes appear in **bold**. Here is an example: "To create a new visualization, click on the **Visualize** tab in the left-hand menu. Click **Create visualization** to start creating a new visualization."

> **Tips or important notes**
> Appear like this.

Get in touch

Feedback from our readers is always welcome.

General feedback: If you have questions about any aspect of this book, email us at customercare@packtpub.com and mention the book title in the subject of your message.

Errata: Although we have taken every care to ensure the accuracy of our content, mistakes do happen. If you have found a mistake in this book, we would be grateful if you would report this to us. Please visit www.packtpub.com/support/errata and fill in the form.

Piracy: If you come across any illegal copies of our works in any form on the internet, we would be grateful if you would provide us with the location address or website name. Please contact us at copyright@packt.com with a link to the material.

If you are interested in becoming an author: If there is a topic that you have expertise in and you are interested in either writing or contributing to a book, please visit authors.packtpub.com.

Share Your Thoughts

Once you've read *Cloud Native Software Security Handbook*, we'd love to hear your thoughts! Scan the QR code below to go straight to the Amazon review page for this book and share your feedback.

https://packt.link/r/1837636982

Your review is important to us and the tech community and will help us make sure we're delivering excellent quality content.

Download a free PDF copy of this book

Thanks for purchasing this book!

Do you like to read on the go but are unable to carry your print books everywhere?

Is your eBook purchase not compatible with the device of your choice?

Don't worry, now with every Packt book you get a DRM-free PDF version of that book at no cost.

Read anywhere, any place, on any device. Search, copy, and paste code from your favorite technical books directly into your application.

The perks don't stop there, you can get exclusive access to discounts, newsletters, and great free content in your inbox daily

Follow these simple steps to get the benefits:

1. Scan the QR code or visit the link below

https://packt.link/free-ebook/9781837636983

2. Submit your proof of purchase
3. That's it! We'll send your free PDF and other benefits to your email directly

Part 1:
Understanding Cloud Native Technology and Security

In this part, you will learn about the foundations of cloud-native technologies, how to secure cloud-native systems, and the security considerations involved in cloud-native application development. By the end of this part, you will have a solid understanding of cloud-native technologies and the security challenges associated with them.

This part has the following chapters:

- *Chapter 1, Foundations of Cloud Native*
- *Chapter 2, Cloud Native Systems Security Management*
- *Chapter 3, Cloud Native Application Security*

1
Foundations of Cloud Native

The adoption of cloud-native solutions is expected to surge in the upcoming years, and platforms such as Kubernetes continue to be the dominant players in this field. With this, the demand for cloud-native technologies and professionals will only continue to rise. This includes the crucial role of cloud-native security engineers and administrators in organizations. Let's dive in and begin with the foundations of cloud-native.

This chapter serves as a comprehensive introduction for those who are familiar with public cloud vendors and their offerings but seek to understand how they integrate with vendor-agnostic cloud-native technologies. We will be exploring a few of the plethora of tools and platforms offered by the **Cloud Native Computing Foundation** (**CNCF**) and delving into the tools and strategies used throughout this book, providing a clear understanding of those platforms, their use cases and applications, and deploying them in real time.

In this chapter, we're going to cover the following main topics:

- Understanding the cloud-native world
- Components for building a cloud-native app
- Approach to thinking cloud-native

Understanding the cloud-native world

If you have been in the tech industry for a while, you are probably aware of the buzzword known as *cloud-native*. The more people you ask what it means, chances are, the more varied answers you will receive, and what's bizarre is that all of them would be accurate in their own way. So, why the different answers? Well, the answer is simple – cloud-native technology and the stack is ever evolving, and each engineer, based on the use case of their cloud-native technology, would consider that in of itself to be cloud-native. However, based on the definition set out by the CNCF and my practical experience of using these technologies for the past many years, instead of defining a broader term of cloud-native computing, I would rather define what it means for an application to be cloud-native:

> *"Cloud-native is the architectural style for any application that makes this application cloud-deployable as a loosely coupled formation of singular services that is optimized for automation using DevOps practices."*

Let's delve into understanding what that means in the industry. Cloud-native is an application design style that enables engineers to deploy any software in the cloud as each service. These services are optimized for automation using DevOps practices such as **Continuous Integration and Continuous Deployment (CI/CD)** and **Infrastructure as Code (IaC)**. This approach allows for faster development, testing, and deployment of applications in the cloud, making it easier for organizations to scale and adapt to changing business needs. Additionally, the use of microservices and containerization in cloud-native architecture allows for greater flexibility and resiliency in the event of service failures. Overall, cloud-native architecture is designed to take full advantage of the cloud's capabilities and provide a more efficient and effective way to build and deploy applications.

Why consider using cloud-native architecture?

I have always found the best way to approach any problem is to start with *why*. As for our current endeavor, it is prudent to think about why we would even care about thinking of a different approach to building our applications when we can get away with the current style of development. While you wouldn't be completely wrong, there are some pretty strong arguments to be made otherwise. While we can address the need for this architecture, further for now, we can try contemplating the benefits of development. A few of them are listed as follows:

- **Scalability**: One of the primary benefits of cloud-native architecture is the ability to easily scale applications horizontally and vertically, to meet changing demands. This is particularly important for applications that experience fluctuating levels of traffic as it allows for resources to be allocated in real time, without the need for manual intervention.

- **Flexibility**: Cloud-native architecture also provides greater flexibility in terms of where and how applications are deployed. Applications can be deployed across multiple cloud providers or on-premises, depending on the needs of the organization, including but not limited to the organization's compliance policies, business continuity, disaster recovery playbooks, and more.

- **Cost savings**: Cloud-native architecture can lead to cost savings as well. By taking advantage of the pay-as-you-go pricing model offered by cloud providers, organizations only pay for the resources they use, rather than having to invest in expensive infrastructure upfront. Additionally, the ability to scale resources up and down can help reduce the overall cost of running applications.

- **Improved security**: Cloud-native architecture also offers improved security for applications. Cloud providers typically offer a range of security features, such as encryption (such as AWS KMS, which is used for encryption key management and cryptographic signing) and multi-factor authentication, which can be applied to applications. Additionally, the use of containerization and microservices can help isolate and secure individual components of an application.

- **Faster deployment**: Cloud-native architecture allows for faster deployment of applications. Containerization, for example, allows you to package applications and dependencies together, which can then be easily deployed to a cloud environment. Frameworks such as GitOps and other IaC solutions help significantly reduce the time and effort required to deploy new applications or updates.

- **Improved resilience**: Cloud-native architecture can also help improve the resilience of applications. By using techniques such as load balancing and automatic failover, applications can be designed to continue running even in the event of a failure. This helps ensure that applications remain available to users, even in the event of disruption.

- **Better performance**: Cloud-native architecture can lead to better performance for applications. By using cloud providers' global networks, applications can be deployed closer to users, reducing latency and improving the overall user experience. Additionally, the use of containerization and microservices can help improve the performance of the individual components of an application.

- **Improved collaboration**: Cloud-native architecture can also improve collaboration among developers. By using cloud-based development tools and platforms, developers can work together more easily and efficiently, regardless of their location. Additionally, the use of containerization and microservices can help promote collaboration among teams by breaking down applications into smaller, more manageable components.

- **Better monitoring**: Cloud-native architecture can also enable better monitoring of applications. Cloud providers typically offer a range of monitoring tools, such as real-time metrics and log analysis, that can be used to track the performance and usage of applications. This can help organizations quickly identify and resolve any issues that may arise.

- **Better business outcomes**: All the aforementioned benefits can lead to better business outcomes. Cloud-native architecture can help organizations deploy new applications, improve the performance and availability of existing applications, and reduce the overall cost of running applications quickly and easily. This can help organizations stay competitive, improve customer satisfaction, and achieve their business goals.

Essentially, there is no silver bullet when it comes to architecting cloud-native applications – the method of architecture heavily depends on the primal stage of defining factors of the application use cases, such as the following:

- **Scalability requirements**: How much traffic and usage is the application expected to handle and how quickly does it need to scale to meet changing demands?

- **Performance needs**: What are the performance requirements of the application and how do they impact the architecture?

- **Security considerations**: What level of security is required for the application and how does it impact the architecture?

- **Compliance requirements**: Are there any specific compliance regulations that the application must adhere to and how do they impact the architecture?

- **Deployment considerations**: How and where will the application be deployed? Will it be deployed across multiple cloud providers, availability zones, or on-premises?

- **Resilience and fault-tolerance**: How should the architecture be designed to handle service failures and ensure high availability?

- **Operational requirements**: How should the architecture be designed to facilitate monitoring, logging, tracing, and troubleshooting of the application in production so that compliance policies such as **service-level indicators (SLIs)**, **service-level objectives (SLOs)**, and error budgets can be applied to the telemetry data that's been collected?

- **Cost and budget**: What is the budget for the application and how does it impact the architecture?

- **Future scalability and extensibility**: How should the architecture be designed to allow for future scalability and extensibility of the application?

- **Integration with existing systems**: How should the architecture be designed to integrate with existing systems and data sources?

While we will discuss a few of those factors in detail in the subsequent chapters, it is important to address the problems and identify the pain points that warrant the use of a cloud-native approach and a design architecture to enable more efficient, scalable systems.

Cloud models

Before we sail into understanding the cloud-native model, it is prudent to understand the existing cloud models for deployment. In this book, to understand the different cloud-native deployment models, I will segregate the cloud offering into two categories.

Cloud deployment model

This deployment model explains strategies of cloud infrastructure deployment from the perspective of the cloud architecture used within the organization and the type of cloud offering that the organization chooses for deployment.

Public cloud

The public cloud is a cloud deployment model in which resources and services are made available to the public over the internet. This includes a wide range of services, such as computing power, storage, and software applications. Public cloud providers, such as **Amazon Web Services (AWS)**, **Microsoft Azure**, and **Google Cloud Platform (GCP)**, own and operate the infrastructure and make it available to customers over the internet. Public cloud providers offer a range of services, including **Infrastructure as a Service (IaaS)**, **Platform as a Service (PaaS)**, and **Software as a Service (SaaS)**, which can be used on a pay-as-you-go basis.

Advantages of the public cloud include flexibility and scalability, as well as cost savings, as customers only pay for the resources they use and do not need to invest in and maintain their infrastructure. Public cloud providers also handle the maintenance and updates/upgrades of the infrastructure, which can free up IT staff to focus on other tasks. Additionally, public clouds are known for providing a global reach, with multiple locations and availability zones, which can help with disaster recovery and business continuity.

While the public cloud offers many advantages, there are also a few potential disadvantages to consider:

- **Security concerns**: Public cloud providers are responsible for securing the infrastructure, but customers are responsible for securing their data and applications. This can create security gaps, especially if customers do not have the necessary expertise or resources to properly secure their data and applications.

- **Limited control and customization**: Public cloud providers offer a wide range of services and features, but customers may not have the same level of control and customization as they would with their own on-premises infrastructure.

- **Vendor lock-in**: Public cloud providers may use proprietary technologies, which can make it difficult and costly for customers to switch to a different provider if they are not satisfied with the service or if their needs change. The operational cost may also rise significantly if the cloud vendor decides to increase the cost of their services, which is difficult to counter in this scenario.

- **Dependence on internet connectivity**: Public cloud services are provided over the internet, which means that customers must have a reliable internet connection to access their data and applications. This can be an issue in areas with limited or unreliable internet connectivity.

- **Compliance**: Public cloud providers may not be able to meet the compliance and regulatory requirements of certain industries, such as healthcare and finance, which may prohibit the use of public cloud services.

- **Data sovereignty**: Some organizations may have data sovereignty requirements that prohibit them from storing their data outside of their own country, and therefore may not be able to use public cloud services.

It's important to carefully evaluate your organization's specific needs and constraints, and weigh them against the benefits of public cloud, before deciding to use public cloud services.

Private cloud

A private cloud is a cloud deployment model in which resources and services are made available only to a specific organization or group of users and are typically operated on-premises or within a dedicated data center. Private clouds are often built using the same technologies as public clouds, such as virtualization, but they are not shared with other organizations. This allows for greater control and customization, as well as higher levels of security and compliance.

In a private cloud, an organization can have full control of the infrastructure and can configure and manage it according to its specific needs and requirements. This allows organizations to have a high degree of customization, which can be important for certain applications or workloads.

The advantages of a private cloud include the following:

- **Greater control and customization**: An organization has full control over the infrastructure and can configure and manage it to meet its specific needs

- **Improved security**: Since the infrastructure is not shared with other organizations, it can be more secure and better protected against external threats

- **Compliance**: Private clouds can be configured to meet the compliance and regulatory requirements of specific industries, such as healthcare and finance

- **Data sovereignty**: Organizations that have data sovereignty requirements can ensure that their data is stored within their own country

Here are some of the disadvantages of a private cloud:

- **Higher cost**: Building and maintaining a private cloud can be more expensive than using a public cloud as an organization has to invest in and maintain its infrastructure

- **Limited scalability**: A private cloud may not be able to scale as easily as a public cloud, which can be an issue if an organization's needs change

- **Limited expertise**: An organization may not have the same level of expertise and resources as a public cloud provider, which can make it more difficult to properly maintain and update the infrastructure

It's important to carefully evaluate the specific needs and constraints of an organization before deciding to use private cloud services.

Hybrid cloud

A hybrid cloud is a combination of public and private clouds, where sensitive data and workloads are kept on-premises or in a private cloud, while less sensitive data and workloads are in a public cloud. This approach allows organizations to take advantage of the benefits of both public and private clouds while minimizing the risks and costs associated with each.

With hybrid cloud, organizations can use public cloud services, such as IaaS and SaaS, to handle non-sensitive workloads, such as web-facing applications and testing environments. At the same time, they can keep sensitive data and workloads, such as financial data or customer data, on-premises or in a private cloud, where they have more control and security.

Here are some of the advantages of a hybrid cloud:

- **Flexibility**: Organizations can use the best cloud services for each workload, which can help improve cost-efficiency and performance
- **Improved security**: Organizations can keep sensitive data and workloads on-premises or in a private cloud, where they have more control and security
- **Compliance**: Organizations can use public cloud services to handle non-sensitive workloads while keeping sensitive data and workloads on-premises or in a private cloud to meet compliance and regulatory requirements
- **Data sovereignty**: Organizations can store sensitive data on-premises or in a private cloud to meet data sovereignty requirements

Disadvantages of a hybrid cloud include the following:

- **Complexity**: Managing a hybrid cloud environment can be more complex than managing a public or private cloud, as organizations need to integrate and manage multiple cloud services
- **Limited scalability**: A hybrid cloud may not be able to scale as easily as a public cloud, which can be an issue if an organization's needs change
- **Limited expertise**: An organization may not have the same level of expertise and resources as a public cloud provider, which can make it more difficult to properly maintain and update the infrastructure
- **Hybrid cloud latency**: If an application in one environment is communicating with a service in another cloud environment, there's a high chance for a bottleneck to be created due to the higher latency of one of the services, leading to increasing the overall latency of the applications

It's important to note that a hybrid cloud environment requires a good level of coordination and communication between the different parts of the organization, as well as with the different cloud providers, to ensure that the different services and data are properly integrated and secured.

Multi-cloud

Multi-cloud is a deployment model in which an organization uses multiple cloud services from different providers, rather than relying on a single provider. By using multiple cloud services, organizations can avoid vendor lock-in, improve resilience, and take advantage of the best features and pricing from different providers.

For instance, an organization might use AWS for its computing needs, Microsoft Azure for its storage needs, and GCP for its big data analytics needs. Each of these providers offers different services and features that are better suited to certain workloads and use cases, and by using multiple providers, an organization can select the best provider for each workload.

Let's look at some of the advantages of the multi-cloud model:

- **Avoid vendor lock-in**: By using multiple cloud services, organizations can avoid becoming too dependent on a single provider, which can be a problem if that provider raises prices or experiences service disruptions

- **Improved resilience**: By using multiple cloud services, organizations can improve their resilience to service disruptions or outages as they can fail over to a different provider if one provider experiences an outage

- **Best features and pricing**: By using multiple cloud services, organizations can take advantage of the best features and pricing from different providers, which can help improve cost-efficiency and performance

- **Flexibility**: Multi-cloud deployment allows organizations to pick and choose the services that best fit their needs, rather than being limited to the services offered by a single provider

The disadvantages of the multi-cloud model include the following:

- **Complexity**: Managing multiple cloud services from different providers can be more complex than managing a single provider as organizations need to integrate and manage multiple cloud services.

- **Limited scalability**: A multi-cloud environment may not be able to scale as easily as a single-cloud environment, which can be an issue if an organization's needs change.

- **Limited expertise**: An organization may not have the same level of expertise and resources as a public cloud provider, which can make it more difficult to properly maintain and update the infrastructure.

- **Higher costs**: Managing multiple cloud services from different providers can be more expensive than using a single provider as organizations need to pay for services and resources from multiple providers. Also, the organization would have to hire multiple engineers that had expertise across all cloud vendors.

It's important for organizations to carefully evaluate their specific needs and constraints, and weigh them against the benefits of multi-cloud, before deciding to use multi-cloud services.

Community cloud

A community cloud is a type of private cloud that is shared by a group of organizations that has similar requirements and concerns. This type of cloud is typically owned, operated, and managed by a third-party provider, and is used by a specific community, such as a group of businesses in a particular industry or a group of government agencies.

Community cloud is a way for organizations to share the costs and benefits of a private cloud infrastructure while maintaining control over their data and applications. For example, a group of healthcare providers may set up a community cloud to share electronic medical records and other healthcare-related data and applications.

The advantages of a community cloud include the following:

- **Cost savings**: Organizations can share the costs of building and maintaining a private cloud infrastructure, which can help reduce costs

- **Specialized resources and expertise**: Community clouds are typically managed by third-party providers that have specialized resources and expertise, which can help improve performance and security

- **Compliance**: Community clouds can be configured to meet the compliance and regulatory requirements of specific industries, such as healthcare and finance

- **Data sovereignty**: Organizations that have data sovereignty requirements can ensure that their data is stored within their own country

Let's look at some of the disadvantages of a community cloud:

- **Limited control and customization**: Organizations may not have the same level of control and customization as they would with their own on-premises infrastructure

- **Security concerns**: Organizations are responsible for securing their data and applications, but they may not have the necessary expertise or resources to properly secure their data and applications

- **Limited scalability**: A community cloud may not be able to scale as easily as a public cloud, which can be an issue if an organization's needs change

- **Limited expertise**: An organization may not have the same level of expertise and resources as a public cloud provider, which can make it more difficult to properly maintain and update the infrastructure

It's important for organizations to carefully evaluate their specific needs and constraints, and weigh them against the benefits of community cloud, before deciding to use community cloud services. Additionally, it's important for organizations using a community cloud to establish clear governance and service-level agreements with other members of the community to ensure smooth operation and prevent conflicts.

> **Important note**
>
> Mostly within organizations in the industry, you would observe a multi-cloud architecture. A part of that reason is that each cloud vendor delivers a particular service in a more efficient way that fits the use case of the application. For those reasons, it is very important to avoid vendor lock-in. This is only feasible if the application is developed in a cloud-native way.

Cloud computing service categories

Cloud computing service categories refer to different levels of abstraction and control over the underlying infrastructure, and they provide different types of services and capabilities. These can be seen in the following diagram:

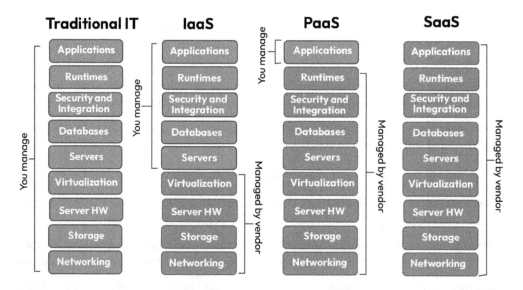

Figure 1.1 – Cloud service model

Let's take a closer look.

IaaS

IaaS is a cloud computing service category that provides virtualized computing resources over the internet. IaaS providers offer a range of services, including servers, storage, and networking, which can be rented on demand, rather than you having to build and maintain the infrastructure in-house. IaaS providers typically use virtualization technology to create a pool of resources that can be used by multiple customers.

IaaS providers typically offer a range of services, including the following:

- **Virtual machines (VMs)**: Customers can rent VMs with specific configurations of CPU, memory, and storage. This allows them to run their operating systems and applications on VMs.

- **Storage**: IaaS providers offer various storage options, such as block storage, object storage, and file storage, that customers can use to store their data.

- **Networking**: IaaS providers offer virtual networks that customers can use to connect their VMs and storage to the internet, as well as to other VMs and services.

The advantages of using IaaS include the following:

- **Cost savings**: Organizations can rent computing resources on demand, rather than building and maintaining their own infrastructure. This can help reduce capital and operational expenses.

- **Scalability**: Organizations can easily scale their computing resources up or down as needed, which can help improve cost-efficiency and performance.

- **Flexibility**: Organizations can choose from a range of VM configurations and storage options, which can help improve performance and security.

- **Improved disaster recovery**: Organizations can use IaaS providers to create backups and replicas of their VMs and storage in different locations, which can help improve disaster recovery and business continuity.

Here are the disadvantages of using IaaS:

- **Limited control**: Organizations may not have the same level of control and customization as they would with their own on-premises infrastructure

- **Security concerns**: Organizations are responsible for securing their VMs and storage, but they may not have the necessary expertise or resources to properly secure their data and applications

PaaS

PaaS is a category of cloud computing services that provides a platform for developers to build, test, and deploy applications without the complexity of managing the underlying infrastructure. PaaS providers typically offer a web server, database, and other tools needed to run an application, such as programming languages, frameworks, and libraries.

PaaS providers typically offer a range of services, such as the following:

- Development tools and environments, such as **integrated development environments** (**IDEs**), version control systems, and debugging tools.

- Deployment and scaling tools, such as automatic load balancing and scaling, and easy rollback and roll-forward of application versions.

- Database services, such as SQL and NoSQL databases, and data storage services.

- Security and compliance features, such as encryption, authentication, and access controls.

- Monitoring and analytics tools, such as logging, performance monitoring, and error reporting.

- Examples of popular PaaS providers include Heroku, AWS Elastic Beanstalk, and Google App Engine. These providers offer a variety of services and tools to help developers quickly and easily build, test, and deploy their applications, without the need to manage the underlying infrastructure. Additionally, PaaS providers often offer usage-based pricing models, making them cost-effective for small and medium-sized businesses.

Let's look at some of the advantages of using PaaS:

- **Faster time to market**: Developers can quickly build, test, and deploy applications without the need to manage the underlying infrastructure, which can help reduce the time to market for new applications.

- **Scalability**: PaaS providers often offer automatic scaling, which allows applications to scale up or down as needed, based on usage or demand

- **Lower costs**: PaaS providers often offer pay-as-you-go pricing models, which can help reduce costs for small and medium-sized businesses

- **Reduced complexity**: PaaS providers often offer pre-configured development environments and tools, which can help reduce the complexity of application development and deployment

- **Improved collaboration**: PaaS providers often offer collaboration tools, such as version control systems, which can help improve collaboration among developers

Here are some of the disadvantages of using PaaS:

- **Limited control**: Developers may not have the same level of control and customization as they would with their own infrastructure or with an IaaS provider

- **Vendor lock-in**: Developers may become reliant on the PaaS provider's tools and services, which can make it difficult to switch providers in the future

- **Compatibility issues**: Applications developed on one PaaS provider may not be compatible with another provider, which can limit flexibility and portability

- **Security concerns**: Developers are responsible for securing their applications and data, but they may not have the necessary expertise or resources to properly secure their applications and data

SaaS

SaaS is a software delivery model in which a software application is hosted by a third-party provider and made available to customers over the internet. SaaS providers manage and maintain the infrastructure, security, and scalability of the software, while customers access the software through a web browser or other remote means.

SaaS applications are typically subscription-based, with customers paying a monthly or annual fee for access. They can be used for a wide range of purposes, including customer relationship management, enterprise resource planning, and human resources management, among others.

SaaS applications are often accessed through a web browser but can also be accessed through mobile apps. They can be used by businesses of all sizes and in a variety of industries, from small start-ups to large enterprise companies. A few examples of applications with SaaS offerings are Jira, Office 365, and Stripe.

The advantages of using SaaS include the following:

- **Easy access**: SaaS applications can be accessed from anywhere with an internet connection, making it convenient for users to access applications from any location or device.

- **Scalability**: SaaS providers often offer automatic scaling, which allows applications to scale up or down as needed, based on usage or demand.

- **Lower costs**: SaaS providers often offer pay-as-you-go pricing models, which can help reduce costs for small and medium-sized businesses. Additionally, SaaS providers are responsible for maintaining the underlying infrastructure and software, which can help reduce IT costs for organizations.

- **Faster implementation**: SaaS applications can be quickly deployed, often within hours or days, without the need for hardware or software installation.

- **Improved collaboration**: SaaS applications often include collaboration tools, such as document sharing and project management tools, which can help improve collaboration among team members.

The disadvantages of using SaaS include the following:

- **Limited control**: Users may not have the same level of control and customization as they would with on-premises software

- **Security concerns**: SaaS providers are responsible for securing the underlying infrastructure and software, but users are responsible for securing their data and applications

- **Dependence on internet connectivity**: SaaS applications require a reliable internet connection, and downtime or slow internet speeds can impact productivity and user satisfaction

- **Data ownership**: Users may have limited control over their data, and there may be limitations on exporting or transferring data to other systems

- **Vendor lock-in**: Users may become reliant on the SaaS provider's applications and services, which can make it difficult to switch providers in the future

Overall, SaaS is a popular and cost-effective way for businesses to access and use software applications without the need to manage and maintain the underlying infrastructure

Approach to thinking cloud-native

As organizations increasingly adopt cloud computing to improve their agility, scalability, and cost-effectiveness, it's becoming critical to think "cloud-native" when designing, building, and deploying applications in the cloud. Cloud-native is an approach that emphasizes the use of cloud computing services, microservices architecture, and containerization to enable applications to be developed and deployed in a more efficient, flexible, and scalable manner.

To help organizations assess their cloud-native capabilities and maturity, the CNCF has developed the **Cloud Native Maturity Model (CNMM)** 2.0. This model provides a framework for organizations to evaluate their cloud-native practices across four levels of maturity: starting out, building momentum, maturing, and leading. Each level includes a set of best practices and capabilities that organizations should strive for as they progress toward cloud-native excellence. By following this model, organizations can ensure that they are building and deploying cloud applications that are optimized for performance, resilience, and scalability, and that can adapt to the dynamic nature of the cloud computing landscape.

CNMM 2.0

CNMM 2.0 is a framework that helps organizations assess and improve their capabilities in developing, deploying, and operating cloud-native applications. It provides a set of best practices and guidelines for designing, building, and running cloud-native applications, along with a set of metrics and indicators to measure an organization's progress and maturity level in implementing these best practices.

The model defines four maturity levels, each representing a different stage of cloud-native maturity – **Initial**, **Managed**, **Proactive**, and **Optimized**. Each level builds on the previous one and has a set of specific characteristics, best practices, and goals that organizations need to achieve to advance to the next level.

CNMM 2.0 is designed to be flexible and adaptable and can be used in any organization, regardless of its size, industry, or cloud provider. It's not limited to a specific cloud service provider.

It's a continuously evolving model that's updated regularly to reflect the latest trends and best practices in cloud-native development and operations.

CNMM 2.0 is a framework that is structured around four maturity levels and four key components. Let's take a look.

Maturity levels

The model defines four maturity levels that organizations can achieve in developing, deploying, and operating cloud-native applications. These levels are displayed in the following diagram:

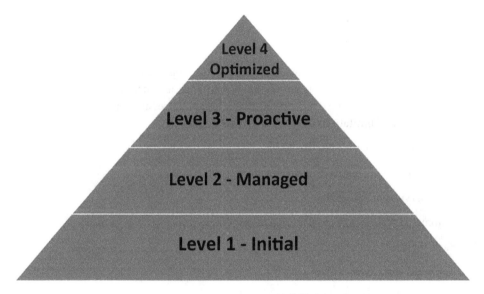

Figure 1.2 – CNMM paradigm

- **Level 1 – Initial**: This level represents an organization's first steps toward cloud-native development and deployment. Organizations at this level may have limited experience with cloud-native technologies and may rely on manual processes and ad hoc solutions.

  Here are the characteristics of this level:

  - Limited use and understanding of cloud-native technologies

  - Monolithic application architecture

  - Limited automation and orchestration

  - Manual scaling and provisioning of resources

  - Limited monitoring and analytics capabilities

  - Basic security measures

Here are the challenges and limitations:

- Difficulty in scaling and managing the application
- A limited understanding of these technologies makes the implementation more error-prone and time-consuming
- Limited ability to respond to changes in demand
- Lack of flexibility and agility
- Limited ability to diagnose and troubleshoot issues
- Increased risk of security breaches
- Limited cost optimization

- **Level 2 – Managed**: This level represents a more mature approach to cloud-native development and deployment, with a focus on automation, governance, and standardization. Organizations at this level have implemented basic cloud-native best practices and have a clear understanding of the benefits and limitations of cloud-native technologies.

  Here are the characteristics of this level:

  - Adoption of cloud-native technologies
  - Microservices architecture
  - Automated scaling and provisioning of resources
  - Basic monitoring and analytics capabilities
  - Improved security measures

  Here are the challenges and limitations:

  - Difficulty in managing the complexity of microservices
  - Limited ability to optimize resources
  - Limited ability to diagnose and troubleshoot issues
  - Limited ability to respond to changes in demand
  - Limited cost optimization

- **Level 3 – Proactive**: This level represents an advanced level of cloud-native maturity, with a focus on continuous improvement, proactive monitoring, and optimization. Organizations at this level have implemented advanced cloud-native best practices and have a deep understanding of the benefits and limitations of cloud-native technologies.

Here are the characteristics of this level:

- Advanced use of cloud-native technologies and practices

- Self-healing systems

- Advanced automation and orchestration

- Advanced monitoring and analytics capabilities

- Advanced security measures

- Optimization of resources

Here are the challenges and limitations:

- Complexity in maintaining and updating automation and orchestration

- Difficulty in keeping up with the fast-paced evolution of cloud-native technologies

- Difficulty in maintaining compliance with security and regulatory requirements

- **Level 4 – Optimized**: This level represents the highest level of cloud-native maturity, with a focus on innovation, experimentation, and optimization. Organizations at this level have implemented leading-edge cloud-native best practices and have a deep understanding of the benefits and limitations of cloud-native technologies.

  Here are the characteristics of this level:

  - Fully optimized use of cloud-native technologies and practices

  - Continuous integration and delivery

  - Predictive analytics and proactive problem resolution

  - Advanced security measures

  - Cost optimization

  Here are the challenges and limitations:

  - Difficulty in keeping up with the latest trends and innovations in cloud-native technologies

  - Difficulty in implementing advanced security measures

  - Difficulty in maintaining cost optimization

Key components

The model defines four key components that organizations need to focus on to achieve different maturity levels. These components are depicted in the following figure:

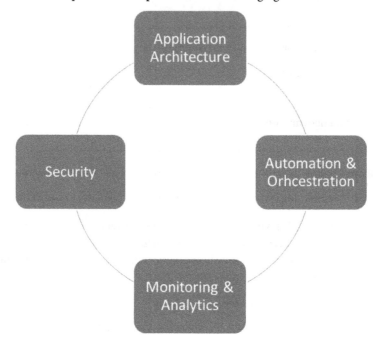

Figure 1.3 – Software deployment component realm

Let's take a look at each component one by one:

- **Application Architecture**

    Application architecture refers to the design and structure of a cloud-native application. It includes characteristics, such as microservices architecture, containerization, cloud agnosticism, and continuous delivery and deployment, all of which are specific to cloud-native applications. These characteristics allow for greater flexibility and scalability in deployment and management on a cloud platform. Best practices for designing and building cloud-native applications include starting small and growing incrementally, designing for failure, using cloud-native services, and leveraging automation.

Here are the characteristics of cloud-native architecture:

- **Microservices architecture**: Cloud-native applications are typically built using a microservices architecture, which involves breaking down a monolithic application into smaller, independent services that can be deployed and managed separately. This allows for greater flexibility and scalability in deployment and management on a cloud platform.

- **Containerization**: Cloud-native applications are often packaged and deployed using containers, which are lightweight, portable, and self-sufficient units that can run consistently across different environments. This allows for greater consistency and ease of deployment across different cloud providers and on-premises environments.

- **Cloud-agnostic**: Cloud-native applications are designed to be cloud-agnostic, meaning they can run on any cloud platform and can easily be moved from one platform to another. This allows for greater flexibility in choosing a cloud provider and in avoiding vendor lock-in.

- **Continuous delivery and deployment**: Cloud-native applications are designed to make use of automated processes and tools for development and operations, such as CI/CD to speed up the development and deployment cycle.

Let's look at the best practices for designing and building cloud-native applications:

- **Starting small and grow incrementally**: Start with a small, simple service and incrementally add more services as needed. This allows for a more manageable and scalable development process.

- **Designing for failure**: Cloud-native applications should be designed to handle failures gracefully, such as by using circuit breakers, load balancers, and self-healing mechanisms.

- **Using cloud-native services**: Utilize the native services provided by the cloud platform, such as databases, message queues, and storage services, to reduce the need for custom infrastructure.

- **Leveraging automation**: Automate as much of the development and deployment process as possible. An example would be to use IaC and CI/CD tools to speed up the development and deployment cycle.

- **Automation and Orchestration**

  Automation and orchestration are key components in cloud-native environments as they help speed up the development and deployment cycle, ensure consistency and reliability in the deployment process, and enable teams to focus on more strategic and value-adding activities. Automation can be achieved by using configuration management tools such as Ansible, Puppet, or Chef to automate the provisioning and configuration of infrastructure, using container orchestration platforms such as Kubernetes, Docker Swarm, or Mesos to automate the deployment, scaling, and management of containers, and using CI/CD tools such as Jenkins, Travis CI, or CircleCI to automate the build, test, and deployment process.

Let's look at the importance of automation in cloud-native environments:

- Automation helps speed up the development and deployment cycle, reducing the time and cost of launching applications to market

- Automation also helps ensure consistency and reliability in the deployment process, reducing the risk of human error

- Automation enables teams to focus on more strategic and value-adding activities

Here are the best practices for automation and orchestration:

- Use an automation tool such as Ansible, Puppet, or Chef to automate the process of provisioning and configuring the infrastructure

- Use container orchestration platforms such as Kubernetes, Docker Swarm, or Mesos to automate the deployment, scaling, and management of containers

- Use CI/CD tools such as Jenkins, Travis CI, or CircleCI to automate the build, test, and deployment process

- Use a service mesh such as Istio or Linkerd to automate how service-to-service communication is managed

- **Monitoring and Analytics**

    Monitoring and analytics are crucial in cloud-native environments as they help ensure the availability and performance of cloud-native applications, provide insights into the behavior and usage of the applications, and help identify and troubleshoot issues. Best practices for monitoring and analytics include using a centralized logging and monitoring solution such as **Elasticsearch, Logstash, and Kibana** (**ELK**). For monitoring metrics and Telemetry, Prometheus and Grafana are commonly used together to collect and visualize system and application-level metrics. Additionally, you can use a distributed tracing system such as Jaeger or Zipkin to trace requests and transactions across microservices and use an **application performance monitoring** (**APM**) solution such as New Relic, AppDynamics, or Datadog to monitor the performance of individual services and transactions.

    Let's look at the importance of monitoring and analytics in cloud-native environments:

    - Monitoring and analytics help ensure the availability and performance of cloud-native applications

    - Monitoring and analytics can provide insights into the behavior and usage of the applications, allowing teams to optimize the applications and make informed decisions

    - Monitoring and analytics also help you identify and troubleshoot issues, allowing teams to resolve problems quickly and effectively

Here are the best practices for monitoring and analytics:

- Use a centralized logging and monitoring solution such as ELK

- Use a distributed tracing system such as Jaeger or Zipkin to trace requests and transactions across microservices

- Use an APM solution such as New Relic, AppDynamics, Prometheus, or Datadog to monitor the performance of individual services and transactions

- Use an A/B testing and experimentation platforms such as Optimizely or Google Optimize to conduct experiments and test new features

- Use a **Business Intelligence (BI)** tool such as Tableau, Looker, or Power BI to analyze data and generate reports

- **Security**

  Security is an essential component in cloud-native environments as applications and data are often spread across multiple cloud providers, making them more vulnerable to attacks. It's also crucial to protect sensitive data, such as personal information, financial data, and intellectual property. Best practices for securing cloud-native applications include using a cloud-native security platform, using a secrets management tool, using a network security solution, using an **identity and access management (IAM)** solution, using encryption to protect data at rest and in transit, and implementing a vulnerability management solution to scan, identify, and remediate vulnerabilities regularly.

  Let's look at the importance of security in cloud-native environments:

  - Security is crucial in a cloud-native environment as applications and data are often spread across multiple cloud providers, making them more vulnerable to attacks

  - Security is also critical in a cloud-native environment to protect sensitive data, such as personal information, financial data, and intellectual property

  - Security is a key part of compliance with regulations, such as the HIPAA, SOC2, and the GDPR

  Here are the best practices for securing cloud-native applications:

  - Use a cloud-native security platform such as Prisma Cloud, Aqua Security, or StackRox to provide security across the entire application life cycle.

  - Use a secrets management tool such as Hashicorp Vault, AWS Secrets Manager, or Google Cloud Secret Manager to securely store and manage sensitive data.

  - Use a network security solution such as AWS Security Groups, Google Cloud Firewall Rules, or Azure Network Security Groups to secure ingress/egress network traffic.

  - Use an IAM solution such as AWS IAM, Google Cloud IAM, or Azure Active Directory to control access to resources and services.

- Use encryption to protect data at rest and in transit. Multiple cloud vendors provide native cryptographic key signing solutions for encryption; they should be regularly revoked and rotated.

- Implement a vulnerability management solution to scan, identify, and remediate vulnerabilities regularly.

CNMM 2.0 provides a set of best practices, metrics, and indicators for each of these four key components, along with a roadmap for organizations to follow as they progress through the four maturity levels. It's designed to be flexible and adaptable, allowing organizations to choose which components and maturity levels they want to focus on, based on their specific needs and goals.

Components of a cloud-native system

As such, multiple projects are a part of the CNCF. For this book, I have agglomerated the platforms and tools that we will use in depth in this book, along with the use case for each platform. However, I strongly recommend that you check out a lot of the others at `https://landscape.cncf.io`:

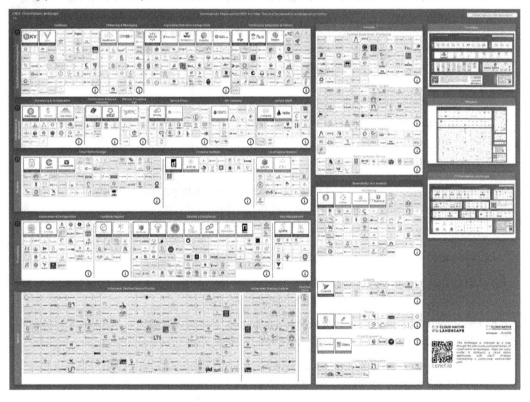

Figure 1.4 – CNCF platform landscape

We will be looking at tools from the following categories:

- Orchestration

- Application development

- Monitoring

- Logging

- Tracing

- Container registries

- Storage and databases

- Runtimes

- Service discoveries and service meshes

- Service proxy

- Security

- Streaming

- Messaging

> **Important note**
>
> You must have a preliminary understanding of how/why these platforms are used in a real system design since the following chapters on threat modeling and secure system design require you to understand how each platform works independently within a cloud-native system, as well as how it integrates with other platforms/tooling/automated processes within the cloud-native system. Also, all the platforms that will be discussed here are cloud-vendor-agnostic.

Orchestration

One of the key projects within the cloud-native space, and the project that we will focus most of our time on, is Kubernetes. Let's take a closer look.

Kubernetes

Kubernetes is a container orchestration system. It allows you to deploy, scale, and manage containerized applications, which are applications that are packaged with all their dependencies, making them more portable and easier to run in different environments.

Kubernetes uses a concept called *pods*, which are the smallest and simplest units in the Kubernetes object model that you can create or deploy. Each pod represents a single instance of a running process in your application. Multiple pods can be grouped to form a higher-level structure called a *ReplicaSet*, which ensures that a specified number of replicas of the pod are running at any given time.

Furthermore, Kubernetes also provides a feature called Services, which allows you to expose your pods to external traffic. It also provides a feature called Ingress, which allows you to route external traffic to multiple services based on the URL path.

Additionally, Kubernetes provides advanced features, such as automatic rolling updates, self-healing, and automatic scaling, which makes it easy to manage and maintain a large number of containers, with some limitations on the number of pods and nodes.

Overall, Kubernetes provides a powerful and flexible platform for deploying and managing containerized applications at scale, making it easier to run, scale, and maintain applications in a production environment.

Monitoring

Multiple tools exist for monitoring code performance, security issues, and other data analytics within the code base, all of which can be leveraged by developers and security engineers. Anecdotally, the following platforms have been widely used in the industry within production environments with the least downtime and the best ease of use.

Prometheus

Prometheus is an open source monitoring and alerting system. It is commonly used for monitoring and alerting on the performance of cloud-native applications and infrastructure.

Prometheus scrapes metrics from different targets, which could be a system, an application, or a piece of infrastructure, and stores them in a time-series database. It also allows users to query and analyze the metrics and set up alerts based on those metrics.

Prometheus is a time-series database that is designed to be highly scalable, and it can handle a large number of metrics, making it suitable for monitoring large-scale systems. It also has a built-in query language called PromQL, which allows users to perform complex queries on the metrics, and a rich set of visualization tools such as Grafana that can be used to display the metrics in a user-friendly way.

Prometheus is also a CNCF project. It is a well-established monitoring tool in the cloud-native ecosystem and is often used in conjunction with other CNCF projects such as Kubernetes.

In summary, Prometheus is an open source monitoring and alerting system that is designed for cloud-native applications and infrastructure. It allows users to scrape metrics from different targets, store them in a time-series database, query and analyze the metrics, and set up alerts based on those metrics. It is also highly scalable and allows for easy integration with other tools and frameworks in the cloud-native ecosystem.

Grafana

Grafana is a powerful tool that allows you to visualize and analyze data in real time. It supports a wide variety of data sources and can be used to create highly customizable dashboards.

One of the key features of Grafana is that it supports Prometheus, a popular open source monitoring and alerting system. Prometheus allows you to collect time-series data from your cloud-native applications and infrastructure, and Grafana can be used to visualize this data in the form of graphs, tables, and other visualizations. This makes it easy to quickly identify trends, patterns, and anomalies in your data and can be used to monitor the health and performance of your systems.

In addition to its visualization capabilities, Grafana also allows you to set up alerts and notifications based on specific thresholds or conditions. For example, you can set up an alert to notify you if the CPU usage of a particular service exceeds a certain threshold, or if the response time of an API exceeds a certain limit. This can help you quickly identify and respond to potential issues before they become critical.

Another of its features is its ability to create a shared dashboard, which allows multiple users to access and interact with the same set of data and visualizations. This can be useful in a team or organization where multiple people are responsible for monitoring and troubleshooting different parts of the infrastructure.

Overall, Grafana is a powerful and flexible tool that can be used to monitor and troubleshoot cloud-native applications and infrastructure.

Logging and tracing

The logical next step after monitoring the deployments is to log the findings for code enhancements and perform trace analysis.

Fluentd

Fluentd is a popular open source data collection tool for the unified logging layer. It allows you to collect, parse, process, and forward logs and events from various sources to different destinations. Fluentd is designed to handle a large volume of data with low memory usage, making it suitable for use in high-scale distributed systems.

Fluentd has a flexible plugin system that allows for easy integration with a wide variety of data sources and outputs. Some common data sources include syslog, HTTP, and in-application logs, while common outputs include Elasticsearch, Kafka, and AWS S3. Fluentd also supports various message formats, such as JSON, MessagePack, and Apache2.

Fluentd can also filter and transform data as it is being collected, which allows you to do things such as drop unimportant events or add additional fields to the log.

It also has a built-in buffering mechanism that helps mitigate the impact of downstream outages and a robust error-handling mechanism that can automatically retry to send the logs in case of failure.

Fluentd's ability to handle a wide variety of data sources and outputs, along with its ability to filter and transform data, makes it a powerful tool for managing and analyzing log data in large-scale distributed systems.

Elasticsearch

Elasticsearch is a distributed, open source search and analytics engine designed for handling large volumes of data. It is often used in cloud-native environments to provide full-text search capabilities and real-time analytics for applications.

One of the main benefits of Elasticsearch for cloud-native environments is its ability to scale horizontally. This means that as the volume of data or the number of users increases, additional nodes can be added to the cluster to handle the load, without requiring any downtime or reconfiguration. This allows Elasticsearch to handle large amounts of data, and still provide low-latency search and analytics capabilities.

Elasticsearch also has built-in support for distributed indexing and searching, which allows data to be partitioned across multiple nodes and searched in parallel, further increasing its ability to handle large volumes of data.

In addition to its scalability, Elasticsearch provides a rich set of features for indexing, searching, and analyzing data. It supports a wide variety of data types, including text, numerical, and date/time fields, and it allows you to perform complex search queries and analytics using its powerful query language, known as the Elasticsearch Query DSL.

Elasticsearch also provides a RESTful API for interacting with the data, making it easy to integrate with other systems and applications. Many popular programming languages have Elasticsearch client libraries that make it even easier to interact with the engine.

Finally, Elasticsearch has a built-in mechanism for handling data replication and sharding, which helps ensure that data is available and searchable even in the event of a node failure. This makes it suitable for use in cloud-native environments where high availability is a requirement.

Overall, Elasticsearch is a powerful tool for managing and analyzing large volumes of data in cloud-native environments, with features such as horizontal scalability, distributed indexing and searching, a rich set of features for indexing, searching, and analyzing data, and built-in support for data replication and sharding.

Kibana

Kibana is a data visualization tool that is commonly used in conjunction with Elasticsearch, a search and analytics engine, to explore, visualize, and analyze data stored in Elasticsearch indices.

In a cloud-native environment, Kibana can be used to visualize and analyze data from various sources, such as logs, metrics, and traces, which is collected and stored in a centralized Elasticsearch cluster. This allows for easy and efficient analysis of data across multiple services and environments in a cloud-based infrastructure.

Kibana can be deployed as a standalone application or as a part of the Elastic Stack, which also includes Elasticsearch and Logstash. It can be run on-premises or in the cloud and can easily be scaled horizontally to handle large amounts of data.

Kibana offers a variety of features for data visualization, such as creating and customizing dashboards, creating and saving visualizations, and creating and managing alerts. Additionally, it provides a user-friendly interface for searching, filtering, and analyzing data stored in Elasticsearch.

In a cloud-native environment, Kibana can easily be deployed as a containerized application using Kubernetes or other container orchestration platforms, allowing you to easily scale and manage the application.

Overall, Kibana is a powerful tool for exploring, visualizing, and analyzing data in a cloud-native environment and can be used to gain valuable insights from data collected from various sources.

Container registries

Within the cloud-native realm, each microservice is deployed within a container. Since they are frequently used within the production environment, it is critical to think about the container registry to be used, and how they're going to be used.

Harbor

Harbor is an open source container registry project that provides a secure and scalable way to store, distribute, and manage container images. It is designed to be a private registry for enterprise usage but can also be used as a public registry. Harbor is built on top of the Docker Distribution open source project and extends it with additional features such as **role-based access control** (**RBAC**), vulnerability scanning, and image replication.

One of the key features of Harbor is its support for multiple projects, which allows you to organize and separate images based on their intended usage or ownership. Each project can have its own set of users and permissions, allowing for fine-grained control over who can access and manage images.

Another important feature of Harbor is its built-in vulnerability scanning capability, which scans images for known vulnerabilities and alerts administrators of any potential risks. This helps ensure that only secure images are deployed in production environments.

Harbor also supports image replication, which allows you to copy images between different Harbor instances, either within the same organization or across different organizations. This can be useful for organizations that have multiple locations or that want to share images with partners.

In terms of deployment, Harbor can be deployed on-premises or in the cloud and can be easily integrated with existing infrastructure and workflows. It also supports integration with other tools such as Kubernetes, Jenkins, and Ansible.

Overall, Harbor is a feature-rich container registry that provides a secure and scalable way to store, distribute, and manage container images and helps ensure the security and compliance of containerized applications.

Service meshes

A service mesh is a vital component in cloud-native environments that helps manage and secure communication between microservices. It provides visibility and control over service-to-service communication, simplifies the deployment of new services, and enhances application reliability and scalability. With a service mesh, organizations can focus on developing and deploying new features rather than worrying about managing network traffic.

Istio

Istio is an open source service mesh that provides a set of security features to secure communication between microservices in a distributed architecture. Some of the key security features of Istio include the following:

- **Mutual TLS authentication**: Istio enables mutual **Transport Layer Security** (TLS) authentication between service instances, which ensures that only authorized services can communicate with each other. This is achieved by automatically generating and managing X.509 certificates for each service instance and using these certificates for mutual authentication.

- **Access control**: Istio provides RBAC for services, which allows for fine-grained control over who can access and manage services. This can be used to enforce security policies based on the identity of the service or the end user.

- **Authorization**: Istio supports service-to-service and end user authentication and authorization using **JSON Web Token** (JWT) and OAuth2 standards. It integrates with external authentication providers such as Auth0, Google, and Microsoft Active Directory to authenticate end users.

- **Auditing**: Istio provides an audit log that records all the requests and responses flowing through the mesh. This can be useful for monitoring and troubleshooting security issues.

- **Data protection**: Istio provides the ability to encrypt payloads between services, as well as to encrypt and decrypt data at rest.

- **Distributed tracing**: Istio provides distributed tracing of service-to-service communication, which allows you to easily identify issues and perform troubleshooting in a distributed microservices architecture.

- **Vulnerability management**: Istio integrates with vulnerability scanners such as Aqua Security and Snyk to automatically detect and alert administrators of any vulnerabilities in the images used for the service.

Overall, Istio provides a comprehensive set of security features that can be used to secure communication between microservices in a distributed architecture. These features include mutual TLS authentication, access control, authorization, auditing, data protection, distributed tracing, and vulnerability management. These features can be easily configured and managed through Istio's control plane, making it simple to secure a microservices environment.

Security

Security provisions have to be applied at multiple layers of the cloud environment, so it is also critical to understand each platform and tool available at our disposal.

Open Policy Agent

Open Policy Agent (**OPA**) is an open source, general-purpose policy engine that can be used to enforce fine-grained, context-aware access control policies across a variety of systems and platforms. It is especially well suited for use in cloud-native environments, where it can be used to secure and govern access to microservices and other distributed systems.

One of the key features of OPA is its ability to evaluate policies against arbitrary data sources. This allows it to make access control decisions based on a wide range of factors, including user identity, system state, and external data. This makes it an ideal tool for implementing complex, dynamic access control policies in cloud-native environments.

Another important feature of OPA is its ability to work with a variety of different policy languages. This makes it easy to integrate with existing systems and tools and allows developers to express policies in the language that best suits their needs.

OPA is often used in conjunction with service meshes and other service orchestration tools to provide fine-grained access control to microservices. It can also be used to secure Kubernetes clusters and other cloud-native infrastructure by enforcing policies at the network level.

In summary, OPA is a powerful and flexible policy engine that can be used to enforce fine-grained, context-aware access control policies across a variety of systems and platforms. It's well suited for use in cloud-native environments, where it can be used to secure and govern access to microservices and other distributed systems.

Falco

Falco is an open source runtime security tool that is designed for use in cloud-native environments, such as Kubernetes clusters. It is used to detect and prevent abnormal behavior in containers, pods, and host systems, and can be integrated with other security tools to provide a comprehensive security solution.

Falco works by monitoring system calls and other kernel-level events in real time and comparing them against a set of predefined rules. These rules can be customized to match the specific requirements of an organization and can be used to detect a wide range of security issues, including privilege escalation, network communications, and file access.

One of the key features of Falco is its ability to detect malicious activity in containers and pods, even if they are running with elevated privileges. This is important in cloud-native environments, where containers and pods are often used to run critical applications and services, and where a security breach can have serious consequences.

Falco can also be used to detect and prevent abnormal behavior on the host system, such as unexpected changes to system files or attempts to access sensitive data. This makes it an effective tool for preventing malicious actors from gaining a foothold in a cloud-native environment.

Falco can be easily integrated with other security tools, such as firewalls, intrusion detection systems, and incident response platforms. It also supports alerting through various channels, such as syslog, email, slack, webhooks, and more.

In summary, Falco is an open source runtime security tool that is designed for use in cloud-native environments. It monitors system calls and other kernel-level events in real time and compares them against a set of predefined rules. This allows it to detect and prevent abnormal behavior in containers, pods, and host systems, making it an effective tool for securing cloud-native applications and services.

Calico

Calico is an open source networking and security solution that can be used to secure Kubernetes clusters. It is built on top of the Kubernetes API and provides a set of operators that can be used to manage and enforce network policies within a cluster.

One of the key security use cases for Calico is network segmentation. Calico allows administrators to create and enforce fine-grained network policies that segment a cluster into different security zones. This can be used to isolate sensitive workloads from less-trusted workloads and prevent unauthorized communication between different parts of a cluster.

Another security use case for Calico is the ability to control traffic flow within a cluster. Calico allows administrators to create and enforce policies that govern the flow of traffic between different pods and services. This can be used to implement micro-segmentation, which limits the attack surface of a cluster by restricting the communication between vulnerable workloads and the external environment.

Calico also provides a feature called *Global Network Policy*, which allows you to define network policies that span multiple clusters and namespaces, enabling you to secure your multi-cluster and multi-cloud deployments.

Calico also supports integration with various service meshes such as Istio, enabling you to secure your service-to-service communication in a more fine-grained way.

In summary, Calico is an open source networking and security solution that can be used to secure Kubernetes clusters. It provides a set of operators that can be used to manage and enforce network policies within a cluster, which can be used for network segmentation, traffic flow control, and securing multi-cluster and multi-cloud deployments. Additionally, it integrates with service meshes to provide more fine-grained service-to-service communication security.

Kyverno

Kyverno is an open source Kubernetes policy engine that allows administrators to define, validate, and enforce policies for their clusters. It provides a set of operators that can be used to manage and enforce policies for Kubernetes resources, such as pods, services, and namespaces.

One of the key security use cases for Kyverno is to enforce security best practices across a cluster. Kyverno allows administrators to define policies that ensure that all resources in a cluster comply with a set of security standards. This can be used to ensure that all pods, services, and namespaces are configured with the appropriate security settings, such as appropriate service accounts, resource limits, and labels.

Another security use case for Kyverno is to provide automated remediation of security issues. Kyverno allows administrators to define policies that automatically remediate security issues when they are detected. This can be used to automatically patch vulnerabilities, rotate secrets, and reconfigure resources so that they comply with security best practices.

Kyverno also provides a feature called *Mutate*, which allows you to make changes to the resource definition before the resource is created or updated. This feature can be used to automatically inject sidecar containers, add labels, and set environment variables.

Kyverno also supports integration with other security tools such as Falco, OPA, and Kube-Bench, allowing you to build a more comprehensive security strategy for your cluster.

In summary, Kyverno is an open source Kubernetes policy engine that allows administrators to define, validate, and enforce policies for their clusters. It provides a set of operators that can be used to manage and enforce policies for Kubernetes resources, such as pods, services, and namespaces. It can be used to enforce security best practices across a cluster, provide automated remediation of security issues, and integrate with other security tools to build a more comprehensive security strategy for a cluster.

Summary

There are multiple tools and platforms available at the disposal of every software engineer within the cloud-native realm. It is important to understand the use case and application of those platforms. When it comes to the model that the product is designed on, you should choose the most efficient and scalable platform.

In this chapter, we tried to provide a clear definition of what we would venture into in this book. I strongly encourage you to read the documentation of the platforms mentioned in this chapter as we will leverage them in this book further and learn about implementing security controls and solutions within any system and application.

With the rise of cloud-native architecture, more companies are adapting to this technique. With that, security engineers and security champions must update their skill sets based on recent updates. In the next chapter, we will be doing a deep dive into understanding secure code development and leveraging the cloud-native approach, as well as a few of the tools discussed in this chapter, to create security solutions for software development.

Quiz

Answer the following questions to test your knowledge of this chapter:

- Why would you want to use cloud-native architecture?
- Why do we care about cloud-native security?
- What are a few components of cloud-native architecture?
- How would you advocate for adopting a cloud-native architecture for your project?

Further readings

To learn more about the topics that were covered in this chapter, take a look at the following resources:

- `https://landscape.cncf.io/`
- `https://community.cncf.io/cncf-cartografos-working-group/`
- `https://kubernetes.io/docs/home/`
- `https://www.redhat.com/en/topics/cloud-native-apps`

2

Cloud Native Systems Security Management

In this chapter, we will be diving into the various security solutions that should be implemented while developing cloud-native systems. Cloud-native systems have become increasingly popular in recent years, but with this increased adoption, security has become a critical concern. This chapter aims to provide you with a comprehensive understanding of the various tools and techniques that can be used to secure cloud-native systems, and how they can be integrated to provide a secure and compliant cloud-native environment.

In this chapter, we're going to cover the following main topics:

- Secure configuration management using OPA
- Secure image management using Harbor

By the end of this chapter, you will be able to implement secure configuration management, secure image management, secure runtime management, secure network management, and Kubernetes admission controllers in their cloud-native systems. You will also be able to integrate various security solutions to provide a secure and compliant cloud-native environment. This is crucial to prevent data breaches, unauthorized access, and other security incidents. The chapter will provide you with a practical guide on how to secure your cloud-native systems, and in turn, it will help you to achieve and maintain compliance and protect your organization's sensitive data and assets.

Technical requirements

The following tools/platforms were installed for this chapter. Please ensure to install the same versions of these tools to run the demos in your local environment:

- Kubernetes v1.27

- Helm v3.11.0

- OPA v0.48

- Harbor v2.7.0

- Clair v4.6.0

- K9s v0.27.2

Here is the GitHub link for the code base: `https://github.com/PacktPublishing/Cloud-Native-Software-Security-Handbook`.

Secure configuration management

In cloud-native systems, configuration management plays a critical role in ensuring the security and compliance of the infrastructure. Configuration management refers to the process of managing, tracking, and controlling changes to a system's configuration. This includes setting up the configurations for the various components of a system, including the operating system, applications, and services, and managing those configurations over time.

In cloud-native systems, configuration management is even more critical due to the dynamic nature of containers and microservices, which require constant updates and changes to configurations. It is important to ensure that these configurations are secure and in compliance with industry standards and best practices.

Open Policy Agent (**OPA**) is a popular tool for secure configuration management in cloud-native environments. OPA is a policy engine that allows organizations to define and enforce policies across their cloud-native infrastructure. It works by providing a centralized location to define policies and make decisions based on these policies.

One of the key benefits of using OPA is that it can be integrated with Kubernetes, the popular container orchestration system, to enforce compliant configurations. This integration allows organizations to automatically enforce policies on their configurations, reducing the risk of human error and increasing the security of their cloud-native systems.

Another key aspect of secure configuration management in cloud-native systems is the management of secrets and sensitive data. This includes sensitive information such as credentials, API keys, and encryption keys. To protect these secrets, it is important to store them in a secure location and manage access to them through proper authentication and authorization mechanisms.

There are several tools available for secure secrets management in cloud-native environments, including HashiCorp Vault, AWS Secrets Manager, and Google Cloud Secret Manager. These tools allow organizations to securely store and manage secrets, ensuring that only authorized users have access to sensitive data (we will look at more secret management tools in *Chapter 3*).

In conclusion, secure configuration management is a crucial aspect of cloud-native systems security and compliance. Organizations can achieve secure configuration management through the use of tools such as OPA and secure secrets management solutions. By implementing these tools and best practices, organizations can reduce the risk of security incidents and ensure that their cloud-native systems are secure and compliant.

Using OPA for secure configuration management

When it comes to policy and compliance, infrastructure changes from company to company and organization to organization. Each team may decide to enforce policies for the infrastructure that they own in a different way, so security teams must define baseline security features that should remain consistent with all the product teams' infrastructure across a single organization.

The following is a list of some of the security techniques that we will be implementing:

- **Requiring encryption for all confidential data**: This policy ensures that all sensitive data, such as passwords and certificates, are encrypted before being stored in the cloud

- **Restricting access to sensitive resources**: This policy can be used to enforce authorization and authentication mechanisms, such as **role-based access control** (**RBAC**), to limit who can access sensitive resources such as databases and secret stores

- **Enforcing resource limits**: This policy can be used to limit the number of resources, such as memory and CPU, that a container can consume, helping to prevent resource exhaustion and improve overall system performance

- **Detecting and preventing vulnerable images**: This policy can be used to automatically scan images for known vulnerabilities and prevent the deployment of images that are known to be vulnerable

- **Monitoring for suspicious activity**: This policy can be used to monitor suspicious activity, such as brute-force attacks, and respond to it in real time

These policies serve as examples of how OPA can be used to enforce compliant configurations in cloud-native environments. By implementing these policies, organizations can ensure that their cloud-native systems are secure and compliant with relevant regulations and standards. Within this section, we will be learning how to leverage OPA and deploy these configurations in a production environment.

> **Important note**
>
> Policies in OPA are written in a language called Rego, which is easy to learn and understand. Additionally, OPA can be used not just for secure configuration management but also for other security use cases, such as API gateway authorization, admission control in Kubernetes, and data protection. This versatility of OPA makes it a valuable tool to secure cloud-native systems. Furthermore, OPA's integration with Kubernetes and other cloud-native technologies makes it a seamless addition to the existing security infrastructure, without requiring any major changes or overhauls.

Requiring encryption for all confidential data

To enforce any policies in your Kubernetes environment, you would first need to install and configure OPA in your cluster. This involves deploying the OPA pod and setting up the necessary resources and permissions.

To install and configure OPA in your Kubernetes cluster, follow these steps:

1. **Deploy the OPA pod**: You can deploy the OPA pod using a YAML file that describes the pod, its resources, and its associated policies. You can use the following YAML file as a reference:

```
apiVersion: apps/v1
kind: Deployment
metadata:
  name: opa
spec:
  replicas: 1
  selector:
    matchLabels:
      app: opa
  template:
    metadata:
      labels:
        app: opa
    spec:
      containers:
      - name: opa
        image: openpolicyagent/opa
        ports:
        - containerPort: 8181
        resources:
          requests:
            memory: "64Mi"
```

```
            CPU: "100m"
         limits:
            memory: "128Mi"
            CPU: "200m"
       command:
       - "run"
       - "--server"
```

Apply the YAML file, using the following command:

```
$ kubectl apply -f opa.yml
```

2. Define the OPA policy in a rego file and store it in a ConfigMap or similar resource in your cluster. This policy will enforce the requirement for encryption of all confidential data:

```
apiVersion: v1
kind: ConfigMap
metadata:
  name: opa-policy
data:
  policy.rego: |
    package opa.kubernetes

    deny[msg] {
      input.review.object.kind == "ConfigMap"
      input.review.object.data.confidential_data
      msg = "Encryption is required for confidential data."
    }
```

To apply the ConfigMap, use the following command:

```
$ kubectl apply -f opa-policy.yml
```

3. Create a Kubernetes admission controller that will enforce the OPA policy. This involves creating a ValidatingWebhookConfiguration object that points to the OPA pod and setting the appropriate rules and parameters:

```
apiVersion: admissionregistration.k8s.io/v1
kind: ValidatingWebhookConfiguration
metadata:
  name: opa-validating-webhook
webhooks:
- name: validate.opa.kubernetes.io
  rules:
  - apiGroups:
    - ""
```

```
        apiVersions:
        - v1
        operations:
        - CREATE
        - UPDATE
        resources:
        - configmaps
      clientConfig:
        caBundle: ""
        service:
          name: opa
          namespace: default
          path: "/v1/data/kubernetes/validate"
```

To apply the Webhook configuration, use the following command:

```
$ kubectl apply -f opa-validating-webhook.yml
```

4. Update your application deployment to include the appropriate annotations and labels, based on your current deployment strategy:

- To ensure that the admission controller can determine which resources are subject to the policy, you will need to add annotations and labels to your deployment manifests.

- For example, you can add an annotation such as `admission.opa.policy=encryption-policy` to indicate that this deployment should be subject to the encryption policy defined in the ConfigMap.

- You can also add labels such as `app=confidential` to indicate that the resources in this deployment contain confidential data that should be encrypted. In our test application, we would have to edit all the pod deployment files to include the annotation, which would look something like this:

```
apiVersion: apps/v1
kind: Deployment
metadata:
  name: auth-api
  labels:
    app: auth-api
spec:
  replicas: 1
  selector:
    matchLabels:
      app: auth-api
  template:
    metadata:
```

```
      labels:
        app: auth-api
      annotations:
        admission.kubernetes.io/webhook: "validation.example.
com/my-validating-webhook-config"
    spec:
      containers:
      - name: auth-api
        image: mihirshah99/auth:latest
        resources:
          limits:
            memory: 512Mi
            cpu: "1"
          requests:
            memory: 256Mi
            cpu: "0.2"
        env:
        - name: AUTH_API_PORT
          valueFrom:
            configMapKeyRef:
              name: auth-api
              key: AUTH_API_PORT
        - name: USERS_API_ADDRESS
          valueFrom:
            configMapKeyRef:
              name: auth-api
              key: USERS_API_ADDRESS
        - name: JWT_SECRET
          valueFrom:
            secretKeyRef:
              name: auth-api
              key: JWT_SECRET
        ports:
        - containerPort: 8081
```

5. Repeat the process, until *step 4*, for all other resources and objects that are deployed in your production cluster:

    -   To ensure that all confidential data in your cluster is encrypted, you will need to repeat this process for all other resources and objects that require encryption

    -   You will need to add the appropriate annotations and labels to each resource and verify that the admission controller and OPA policy work correctly

6.   Regularly monitor and audit your cluster, once you've deployed the OPA file we created in the previous steps:

   • To ensure that the policy is enforced and that all confidential data remains properly encrypted, you will need to regularly monitor and audit your cluster

   • You can use tools such as OPA's audit logs and Kubernetes events to track policy enforcement and monitor for any security incidents

   • You should also regularly review and update your OPA policies to reflect changes in your infrastructure and ensure that they remain up to date and effective

---

Important note

It is also recommended to integrate OPA with other security tools and solutions in your cluster, such as security scanners and monitoring tools, to provide a comprehensive and unified approach to securing your cloud-native environment. This can help to catch potential security incidents early and to respond quickly and effectively to address any issues that may arise. Since a lot of these changes also require updating all deployment files, using automation tools with OPA also helps ease this process. We will explore some of those automation tools later in the book.

Restricting access to sensitive resources

To restrict access to sensitive resources in your Kubernetes environment, you need to follow these steps:

1.   **Write OPA policies**: Create a file called `policy.rego` to define your policies. In this example, you will restrict access to secrets:

```
package kubernetes.authz

default allow = false

allow {
    input.method == "get"
    input.path = ["api", "v1", "namespaces", namespace,
"secrets"]
    namespace := input.params["namespace"]
}

allow {
    input.method == "list"
    input.path = ["api", "v1", "namespaces", namespace,
"secrets"]
    namespace := input.params["namespace"]
}
```

2. **Deploy the OPA policy**: Create a ConfigMap to store the policy and deploy it to your cluster:

```
apiVersion: v1
kind: ConfigMap
metadata:
  name: opa-policy
data:
  policy.rego: |
    package kubernetes.authz

    default allow = false

    allow {
        input.method == "get"
        input.path = ["api", "v1", "namespaces", namespace,
"secrets"]
        namespace := input.params["namespace"]
    }

    allow {
        input.method == "list"
        input.path = ["api", "v1", "namespaces", namespace,
"secrets"]
        namespace := input.params["namespace"]
    }
```

3. **Use OPA as a Kubernetes admission controller**: Create `ClusterRole` and `ClusterRoleBinding`, which gives OPA the necessary permissions to enforce the policies:

```
apiVersion: rbac.authorization.k8s.io/v1
kind: ClusterRole
metadata:
  name: opa-cluster-role
rules:
  - apiGroups: [""]
    resources: ["configmaps"]
    verbs: ["get", "watch"]
  - apiGroups: ["authentication.k8s.io"]
    resources: ["tokenreviews"]
    verbs: ["create"]

---
apiVersion: rbac.authorization.k8s.io/v1
kind: ClusterRoleBinding
```

```
metadata:
  name: opa-cluster-role-binding
roleRef:
  apiGroup: rbac.authorization.k8s.io
  kind: ClusterRole
  name: opa-cluster-role
subjects:
  - kind: ServiceAccount
    name: opa
    namespace: default
```

To apply the OPA cluster role configuration, use the following command:

```
$ kubectl apply -f opa-rbac.yaml
```

4. **Deploy OPA as a deployment**: Create a deployment that runs OPA in your cluster:

```
apiVersion: apps/v1
kind: Deployment
metadata:
  name: opa
  namespace: default
spec:
  replicas: 1
  selector:
    matchLabels:
      app: opa
  template:
    metadata:
      labels:
        app: opa
```

To apply the deployment, use the following command:

```
$ kubectl apply -f opa-deployment.yaml
```

> **Important note**
>
> To enforce any additional restrictions, you would have to update the rego file for OPA and create new ClusterRoles and ClusterRoleBindings to enforce that OPA policy. It is critically important to have granular access over all objects deployed within a cluster, of which OPA does a brilliant job.

Enforcing resource limits

To enforce resource limits in Kubernetes using OPA, you can follow these steps:

1. **Write OPA policies**: Create a file called `policy.rego` to define your policies. In this example, you will enforce memory and CPU limits for containers:

```
package kubernetes.resource_limits

default allow = false

allow {
    input.kind == "Pod"
    pod := input.object
    containers := [container | container := pod.spec.
containers[_]]
    limits := [container.resources.limits[_]]
    mem_limit := limits[_].memory
    cpu_limit := limits[_].cpu

    mem_limit != ""
    cpu_limit != ""
}
```

2. **Deploy the OPA policy**: Create a ConfigMap to store the policy and deploy it to the cluster:

```
apiVersion: v1
kind: ConfigMap
metadata:
  name: opa-policy
data:
  policy.rego: |
    package kubernetes.resource_limits

    default allow = false

    allow {
        input.kind == "Pod"
        pod := input.object
        containers := [container | container := pod.spec.
containers[_]]
        limits := [container.resources.limits[_]]
        mem_limit := limits[_].memory
        cpu_limit := limits[_].cpu
```

```
        mem_limit != ""
        cpu_limit != ""
    }
```

3. **Use OPA as a Kubernetes admission controller**: Create a cluster rule and cluster role binding that gives OPA the necessary permissions to enforce the policies:

```
apiVersion: rbac.authorization.k8s.io/v1
kind: ClusterRole
metadata:
  name: opa-cluster-role
rules:
  - apiGroups: [""]
    resources: ["configmaps"]
    verbs: ["get", "watch"]
  - apiGroups: ["authentication.k8s.io"]
    resources: ["tokenreviews"]
    verbs: ["create"]


---
apiVersion: rbac.authorization.k8s.io/v1
kind: ClusterRoleBinding
metadata:
  name: opa-cluster-role-binding
roleRef:
  apiGroup: rbac.authorization.k8s.io
  kind: ClusterRole
  name: opa-cluster-role
subjects:
  - kind: ServiceAccount
    name: opa
    namespace: default
```

4. **Deploy OPA as a deployment**: Create a deployment that runs OPA in your cluster:

```
apiVersion: apps/v1
kind: Deployment
metadata:
  name: opa
  namespace: default
```

```
spec:
  replicas: 1
  selector:
    matchLabels:
      app: opa
  template:
    metadata:
      labels:
        app: opa
```

5. To enforce resource limits in a specific namespace (in this case, default) and for a specific resource type (in this case, deployments), you can write an over policy rego to the same ConfigMap, as follows:

```
package kubernetes.authz
default allow = false
allow {
  input.kind == "Deployment"
  input.namespace == "default"
  input.spec.template.spec.containers[_].resources.limits[_].
memory <= "512Mi"
  input.spec.template.spec.containers[_].resources.limits[_].cpu
<= "500m"
}
```

This policy checks that the deployment is in the default namespace and limits the memory used by each container in the deployment to 512 Mi and the CPU to 500 m. These are some of the security solutions that can be used using OPA; you can create others based on your organization's compliance requirements. Now that we have deployed the clusters and monitored them, let's do a deep dive into image security.

> **Important note**
>
> Not all containers within a cluster should be subjected to this policy, since a lot of monitoring and audit containers do sometimes require higher memory and CPU usage. The use of an OPA policy for this use case must be fine-grained (another plus point for OPA) and should only be used for containers with a preset function task. Enforcing this policy also protects your infrastructure from crypto mining and other denial-of-wallet attacks. You can learn more about this here: https://mihirshah99.medium.com/stock-manipulation-by-exploiting-serverless-cloud-services-9cea2aa8c75e.

Secure image management

As organizations adopt containers and container orchestration platforms such as Kubernetes, it becomes increasingly important to secure the container images used to deploy applications. Containers, by design, run isolated from the host and other containers, making them a secure way to deploy applications. However, the security of a container is only as strong as the security of the underlying image it is built from. Vulnerabilities in the base image or the software packages included in the image can compromise the security of the entire container and the host.

Therefore, organizations must implement secure image management practices to ensure the security of their container deployments. This involves identifying and mitigating vulnerabilities in the images used to build containers, and ensuring that the images deployed to production are secure and up to date. In this section, we will discuss the importance of secure image management and the best practices to secure container images.

Why care about image security?

An image repository is a central store for all images that are fetched during container runtime, so it is vital to look into image security even before an image gets deployed, for the following reasons:

- **Protecting against known vulnerabilities**: One of the biggest security concerns with containers is the use of vulnerable images. Base images or application images that contain known vulnerabilities can expose a container and the host to attack. Vulnerabilities can range from known exploits for operating systems or applications to misconfigurations that leave the container or host open to attack. By implementing secure image management practices, organizations can scan images for known vulnerabilities and prevent the deployment of images that contain them.

- **Ensuring compliance**: Organizations are also tasked with complying with various regulations and standards such as PCI-DSS and HIPAA, and security certifications such as FedRAMP, which require the secure deployment and management of applications. By implementing secure image management practices, organizations can ensure that the images they use to deploy applications comply with these regulations and standards, reducing the risk of non-compliance and associated penalties.

- **Maintaining the integrity of images**: As images are pulled and deployed, it is important to ensure the integrity of the images used. Tampered or modified images can contain malware or other malicious code, compromising the security of a host and other containers. By implementing secure image management practices, organizations can ensure that the images they use have not been tampered with and are secure.

- **Keeping images up to date**: Vulnerabilities in software packages and base images can be fixed with updates and patches. By implementing secure image management practices, organizations can ensure that the images they use are up to date, reducing the risk of vulnerabilities and attacks.

Best practices for secure image management

Often, organizations are bound to follow best practices as part of an SLA or a security model defined by the compliance teams. The following are a few of the best practices that are generically followed across all organizations:

- **Scanning images for vulnerabilities**: One of the first steps in implementing secure image management is to scan images for vulnerabilities. This involves using a vulnerability scanner to identify known vulnerabilities in the images used to build containers. Some popular vulnerability scanners include Clair, Aqua Security, and Snyk.

- **Integrating with container registries**: It is best practice to integrate vulnerability scanning with the container registry used to store and manage images. This ensures that images are scanned automatically as they are pushed to the registry, reducing the risk of deploying vulnerable images.

- **Building secure images**: Organizations should build their images using a secure and reproducible process, including the use of base images from trusted sources, updating images regularly, and adopting best practices for software package management.

- **Using signatures and image digests**: Images should be signed and their digests recorded to ensure the authenticity and integrity of the images used. This helps to prevent the deployment of tampered images and reduces the risk of attack. Also, organizations should always ensure that the images used are the latest and have automated checks in place to pull the latest image from the vendor.

To facilitate implementing the preceding best practices, we will leverage cloud-native tools such as Harbor, Snyk, and Clair for their respective use cases. Let's consider the pros and cons of using each tool next:

- Harbor is a cloud-native registry that stores, signs, and scans container images for vulnerabilities

- Snyk is a vulnerability scanner that identifies known vulnerabilities in the images used to build containers

- Clair is an open source project for the static analysis of vulnerabilities in application containers (including Docker) and images.

By combining these tools, organizations can implement a secure and efficient process to manage the images used in their cloud-native environments, reducing the risk of deploying vulnerable images and ensuring the integrity and authenticity of the images used.

Let us first begin by understanding how we would perform static analysis for code vulnerabilities on-premises – Clair could be a great tool for local scanning and integration with CI/CD pipelines on-premises (there's more on this in *Chapter 8*).

Clair

Setting up a local environment for image scanning can be a complex and time-consuming process if you follow the traditional method. The following guide provides a simpler method to set up a lab environment for image scanning, using Docker Compose:

1. Clone or download the configuration YAML files from GitHub.
2. The configuration files will create a private network, and all containers will be added to the network, making communication between containers easier.
3. Pull the image you want to scan and tag it with a desired name.
4. Push the tagged image to your local private registry (Harbor in this case).
5. Use Clair to scan the images against the **Common Vulnerabilities and Exposures (CVE)** list in the PostgreSQL database.
6. If any vulnerabilities match CVE, they will be copied and stored in persistent volume storage.

These are the components of the local environment:

- **Testing image**: A container that runs the testing image from your private registry.
- **Clair**: A container that runs Clair within CoreOS, responsible for scanning images.
- **PostgreSQL**: A container that runs a PostgreSQL image, storing all the CVEs. The CVEs must be updated manually at this time.

A high-level architecture of the test design would be as follows:

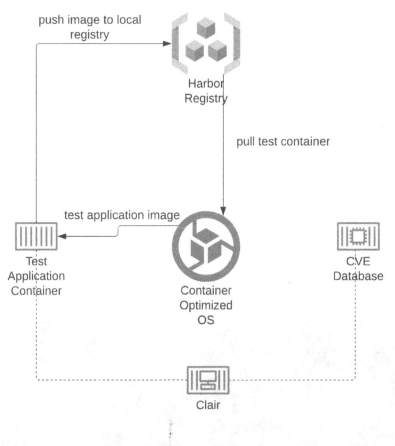

Figure 2.1 – A Clair deployment diagram

> **Important note**
> While the setup is demonstrated using Docker Compose, it can also be deployed within Kubernetes for more advanced use cases.

After cloning the repository with the configuration YAML files, follow these steps:

1. cd into the downloaded folder and execute the following command:

    ```
    $ docker-compose up -d
    ```

2. Install Klar. Klar is a CLI tool integration for Clair; at the time of writing, Klar supports integration with Amazon Container Registry, Google Container Registry, and Docker Registry. Since Clair is, after all, an API-driven vulnerability scanning tool, it is our responsibility to feed the containers. The installation procedure for Klar can be found here: https://github. com/optiopay/klar.

3. Verify that the docker-compose network is up and running; you can also verify this by doing exec into the container, doing a host discovery, and finding other hosts (containers) within the same subnet.

4. The next step is pulling the image from a third-party repository, tagging it locally, and pushing it to our local registry. We will use Docker Hub for this example. Run the following commands:

    ```
    $ docker pull nginx:latest
    $ docker tag nginx localhost:5000/nginx-test
    $ docker push localhost:5000/nginx-test
    ```

    You should be able to see a successful container spin-up, as shown here:

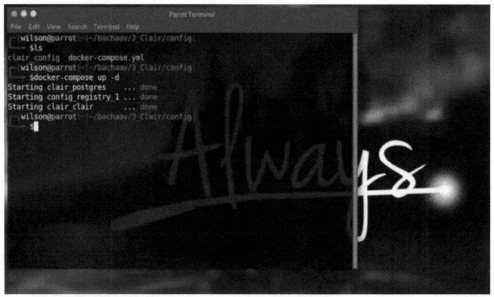

Figure 2.2 – Clair up and running

For this example, we will use the nginx image and try to find vulnerabilities in it.

5. We now must run the test by using Klar, which employs Clair to test the container against the vulnerability database. Set the following local environment variables, since this is used by Clair at the time of execution:

```
$ CLAIR_ADDR=http://localhost:6060 CLAIR_OUTPUT=Low CLAIR_
THRESHOLD=10 REGISTRY_INSECURE=TRUE klar localhost:5000/nginx-
test
```

Let's try and understand the preceding variables better:

- `CLAIR_ADDR`: This variable holds the endpoint location for Clair to run upon

- `CLAIR_OUTPUT=Low`: This informs Clair to output all the CVE matches from the database, including *Low*, *Medium*, *High*, or *Critical*

- `CLAIR_THRESHOLD=10 REGISTRY_INSECURE=TRUE`: This is a mandatory variable that needs to be declared to run Clair locally

All the variables are then accessed by Klar before passing them as arguments to run Clair. The output will be similar to what is displayed in the following screenshot:

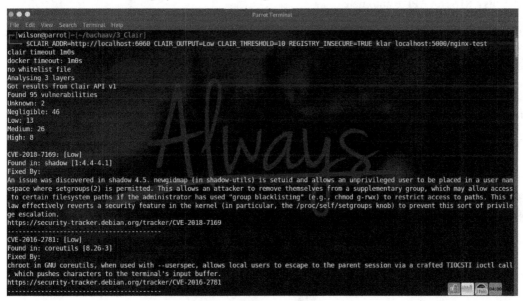

Figure 2.3 – A Clair vulnerability report

This log file can also be saved and exported to multiple other **software composition analysis (SCA)** tools for further investigations – for example, Mend and Checkmarx SCA.

Harbor

Harbor is an open source registry to store, distribute, and manage Docker images. It is designed to provide enterprise-level security features, such as user authentication and image vulnerability scanning, for use in production environments. Some of the benefits and use cases of Harbor include the following:

- **Security**: Harbor offers a secure platform to store and manage Docker images, with features such as RBAC and image vulnerability scanning

- **Compliance**: Harbor provides auditing and reporting capabilities to help organizations comply with regulatory requirements and industry standards

- **Scalability**: Harbor is designed to scale with the needs of large organizations and supports high availability and disaster recovery

- **Interoperability**: Harbor is compatible with other Docker-based tools and systems, making it easy to integrate into existing workflows and infrastructures

- **Ease of use**: Harbor provides a user-friendly web interface to manage Docker images, as well as a RESTful API to automate tasks

Overall, Harbor is a valuable tool for organizations that need to manage and secure Docker images in production environments. In this section, we will focus on using Trivy as the built-in image scanning engine within Harbor, using Harbor securely to push Docker images to it, and monitoring Harbor. Let's begin by first understanding the high-level architecture of Harbor, as an image repository.

Harbor's high-level architecture

Understanding high-level architecture can provide deeper knowledge about the ins and outs of tools, and I am also just fascinated by the system design of platforms that are used by a wider audience to solve complex security issues at scale. Harbor is one of those platforms and provides a secure image registry at an enterprise level with enterprise-grade security; hence, I believe it is beneficial to learn how the tool works. Here's a high-level architecture diagram of Harbor:

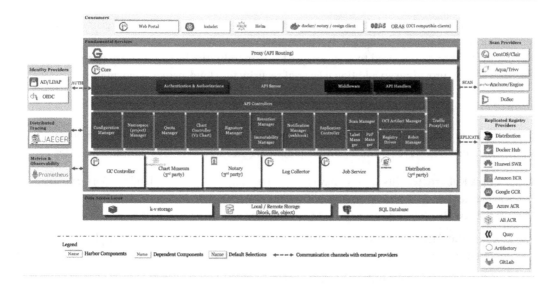

Figure 2.4 – A high-level diagram of Harbor components

The Harbor architecture consists of several components that work together to provide a complete registry solution:

- **Database**: Harbor uses a relational database management system to store metadata information, such as user authentication and authorization, repository information, and image tags.

- **Registry**: The Harbor registry component is responsible for storing Docker images and serving them to users. It uses the Docker distribution open source project to store images.

- **Web UI**: Harbor provides a user-friendly web interface for users to manage their Docker images, including uploading, downloading, and deleting images.

- **Job service**: The job service component is responsible for performing various tasks within Harbor, such as vulnerability scanning, image replication, and image deletion.

- **Token service**: The token service component generates and manages authentication tokens for users accessing the Harbor registry.

- **Notary**: Notary is an open source component that provides content trust and image-signing functionality for Docker images. Harbor integrates with Notary to provide content trust and image-signing capabilities.

- **Trivy**: Trivy is, again, an open source tool to scan images for static vulnerabilities. Harbor uses Trivy's scanning engine to find vulnerabilities within the images stored in the repositories.

Harbor integrates with Kubernetes by providing a secure and scalable Docker image registry solution. Users can store their Docker images in Harbor and access them from their Kubernetes clusters. Additionally, Harbor can be used to scan images for vulnerabilities before they are deployed in a Kubernetes cluster, providing an additional layer of security for cloud-native deployments.

Running Harbor on Kubernetes

Deploying Harbor on Kubernetes can be a complex process, given the intricacies involved in operating a Kubernetes cluster; however, it can be made operational on Kubernetes by following these steps:

1. **Set up a Kubernetes cluster**: You may choose to use any of the major cloud vendors' managed Kubernetes service, or an on-premises installation such as minikube or `docker-desktop`. We will use `docker-desktop` for the installation of Harbor in this chapter.

2. **Install Helm**: Helm is a package manager for Kubernetes that makes it easier to deploy, upgrade, and manage applications.

3. **Add the Helm Chart repository**: Harbor provides a chart repository that contains the necessary manifests to deploy Harbor on the Kubernetes cluster. The repository can be added to Helm by executing the following command:

    ```
    $ helm repo add harbor https://helm.goharbor.io
    ```

4. **Download the Harbor chart**: The Harbor chart can be downloaded using the following command:

    ```
    $ helm fetch harbor/harbor
    ```

5. **Configure the Harbor chart**: The Harbor chart can be configured using a values file that contains the desired configuration values. This file can be created using the following command:

    ```
    $ helm show values harbor/harbor > values.yaml
    ```

    You can find the updated `values.yaml` file in the GitHub repository here: `https://github.com/PacktPublishing/Cloud-Native-Software-Security-Handbook/blob/main/chapter_02/ToDo_app/installation/harbor-values.yaml`.

6. **Create a custom namespace**: It is always a good idea to create a custom namespace for any tool addition within your Kubernetes cluster; this ensures that, within the production cluster, the ops teams have a better triage process to find each deployment. You can create a custom namespace by executing the following command:

    ```
    $ kubectl create namespace my-harbor
    ```

7. **Install the Harbor chart**: The Harbor chart can be installed using the following command:

    ```
    $ helm install my-harbor harbor/harbor -f values.yaml
    ```

8. **Verify the installation**: The installation can be verified by checking the status of the Harbor pods and services using the following command:

```
$ kubectl get pods,services
```

I am using the K9s terminal viewer for Kubernetes, but in any case, you should be able to see the Harbor pods and services up and running, as shown here:

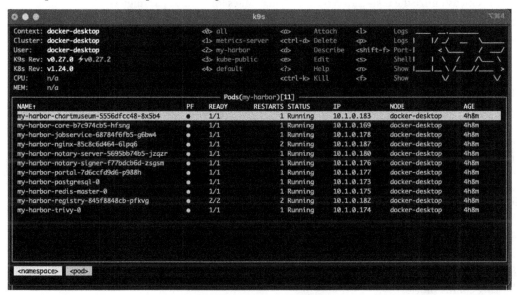

Figure 2.5 – Harbor running pods

In the preceding figure, you can see that I have also installed the nginx pod. This is because we want to access the Harbor UI from the browser. To do that, not only would we need an Ingress Controller to route the incoming requests from our browser to the correct pod within Kubernetes, but we would also need a service for the core-harbor pod. Note that this service would have to be of the type LoadBalancer. You can install ingress-nginx using Helm by performing the following steps:

1. Add the nginx Ingress repository to your Helm chart:

```
$ helm repo add ingress-nginx https://kubernetes.github.io/
ingress-nginx
```

2. Update your Helm repository to include the latest charts:

```
$ helm repo update
```

3. Install the nginx Ingress Controller using Helm with the following command:

```
$ helm install ingress-nginx ingress-nginx/ingress-nginx –
namespace my-harbor
```

> **Important note**
>
> You can customize the installation by specifying values for various parameters – for example, to specify a namespace. In this case, we need to install the Ingress Controller within the same namespace as the Harbor deployment. This is because each namespace follows its networking scheme, and deployments in one namespace do not have any networking access to the deployments in the other namespaces, unless we explicitly define a service with type `ClusterIP` to create a network link.

To access the Harbor UI after successfully installing it in a namespace within your Kubernetes cluster, you need to set up a service to expose the Harbor UI. One way to do this is to create a `LoadBalancer` service in Kubernetes. This service will expose the Harbor UI to the internet by creating an external IP address.

You can use the following command to create a `LoadBalancer` service in Kubernetes:

```
$ kubectl expose deployment my-harbor-harbor --type=LoadBalancer
--port=80 --target-port=8080
```

After creating the `LoadBalancer` service, you can find the external IP address while running the following command:

```
$ kubectl get svc my-harbor-harbor
```

You should see the following services running within the `my-harbor` namespace of your cluster:

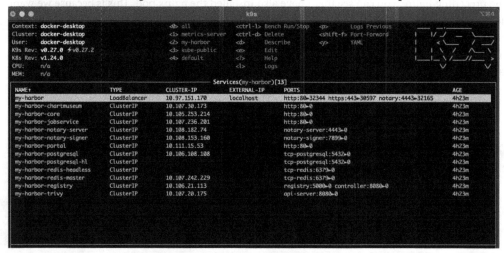

Figure 2.6 – The Harbor service mesh

In my case, the external IP is the localhost. Once you have an external IP address, you can access the Harbor UI by visiting `http://localhost:80`, and you will be greeted by an amazing login page:

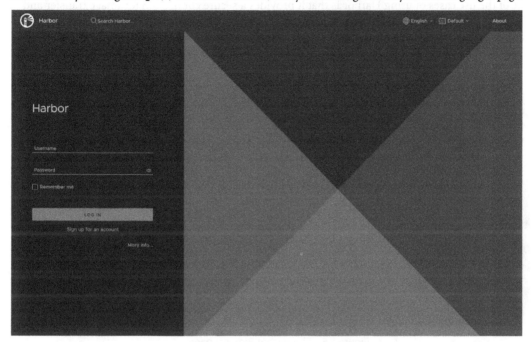

Figure 2.7 – The Harbor portal login

At this point, Harbor is up and running, and you can now log in with the default login credentials as defined in the Helm `values.yaml` file – `admin` for the username and `harbor12345` for the password at the time of writing. Please refer to the Harbor documentation to verify your installation, as it might change depending on the various methods that you might want to install this.

Creating an HTTPS connection for the repository

One of the use cases of using Harbor is that it provides a secure connection for image transfer and storage. To enable that, we need to provide our Harbor server (Kubernetes pod deployment) with an x509 certificate. This can be provisioned by following these steps:

1.   Go to **Projects**, and click on **library**:

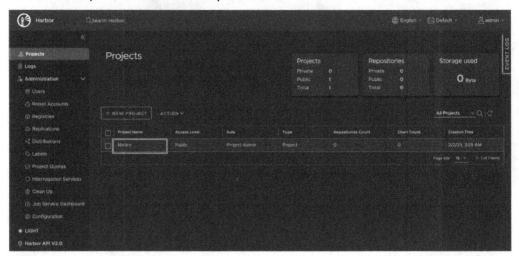

Figure 2.8 – The Harbor Projects page

2.   Next, click on the **REGISTRY CERTIFICATE** button to download the `ca.crt` file. This is a self-signed certificate generated by Harbor and can be used to enable an HTTPS connection with the Kubernetes pod:

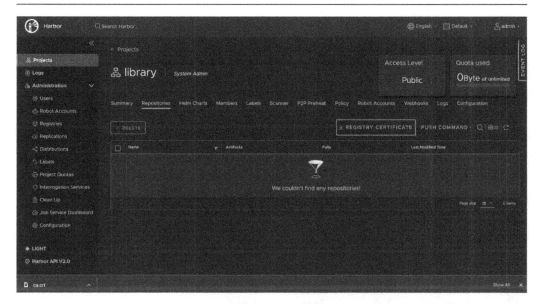

Figure 2.9 – Harbor registry certificate verification

Once the file has been downloaded, the certificate should be added to your system's trust zone; follow this guide (`https://docs.docker.com/registry/insecure/#use-self-signed-certificates`) to understand your system-specific process to install the certificate. Once this has been configured, you should now be able to connect to the Harbor UI over HTTPS, and any images pushed to the registry will also be over HTTPS.

You need to restart your Docker daemon so that the certificate gets reevaluated by the Harbor server. Once all the pods and services are up and running, log in to your Docker client using the credentials for Harbor:

```
$ docker login -p admin https://localhost
Password:
Login Succeeded
```

Once we have added the certificate and ensured that all communication with our local Harbor registry is encrypted and secure, we can start scanning the images for vulnerabilities.

Scanning for vulnerabilities in images

Similar to our previous efforts of scanning vulnerabilities in Clair, we will now scan vulnerabilities within Harbor but with the use of Trivy (Harbor's built-in scanning engine). Perform the following steps to manually scan an image:

1. Push the testing image to the registry using the Docker client:

    I.   We will download the latest image from `nginx: latest`, by running the following command:

    ```
    $ docker pull nginx:latest
    ```

    II.   Tag the downloaded image to our repository using the following command:

    ```
    $ docker tag nginx:latest localhost/library/nginx:latest
    ```

    III. Push this image to the Harbor registry using the following command:

    ```
    $ docker push localhost/library/nginx:latest
    ```

2. Once the image has been pushed, you should be able to view the image name in the **Repositories** tab of the dashboard:

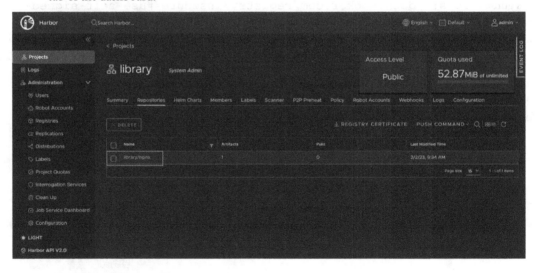

Figure 2.10 – Image scanning using Harbor

3. To enable scanning, we will go to the **Configuration** tab and select the **Automatically scan images on push** option, and then click on **Save**.

4. Once you have verified that you can push the images, you can select the checkbox for the image name, click on **delete** to push the image again, and look at the scan results, by clicking on the name of the image and reading the **Artifacts** page:

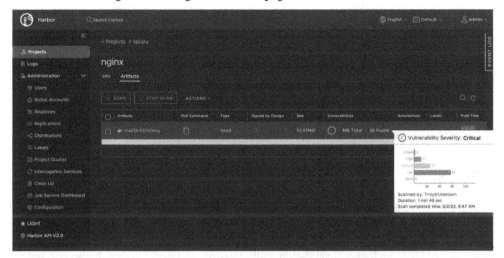

Figure 2.11 – Vulnerability analysis in Harbor

5. By clicking on the **Artifacts** results, you should be able to view the details of each vulnerability detected within the resulting scan. For the latest nginx image, at the time of writing, the following vulnerabilities are detected:

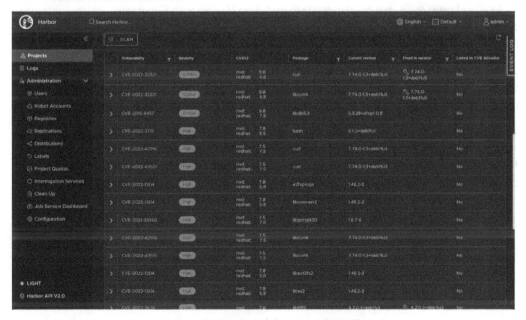

Figure 2.12 – A vulnerability report by Harbor

> **Important note**
>
> The results also inform us whether the image is signed or not; the verification of image signatures is done by Notary (a CNCF-supported tool to verify image signatures). Security engineers can also enable a policy check that prevents image pulls if the images are not signed, or if the scan results for the image are above the set criticality (Low, Medium, High, or Critical).

Harbor allows many other automation and integrations with other third-party commercial tools, such as Snyk and Aqua MicroScanner. We will discuss those in detail in *Chapter 8*. Although the initial setup of Harbor involves a lot of intricacies, it is worth putting the effort in, given the unified interface within the dashboard, secure image transfer, encryption for images at rest, image scanning and signature verification, and many other features bundled within this tool!

Summary

This chapter provided a comprehensive overview of two important tools, OPA and Harbor, and their applications in modern DevOps workflows. OPA is a policy-as-code tool that enables organizations to manage and enforce policy decisions across their infrastructure. The chapter outlined several real-world use cases of OPA, such as API validation, authorization, and data filtering. It also covered the key features of OPA, including its flexible architecture, rich query language, and integrations with various platforms.

Harbor, on the other hand, is a cloud-native registry that provides secure storage, distribution, and management of Docker images. This chapter explained the need for Harbor in modern DevOps workflows and provided a step-by-step guide to installing Harbor with Kubernetes. A demo was also included to show how Harbor can be used to scan for vulnerabilities in Docker images, ensuring that the images used in production are secure and free from security threats.

This chapter also highlighted the synergies between OPA and Harbor. By combining OPA's policy management capabilities with Harbor's security and image management capabilities, organizations can enhance the security and policy enforcement of their infrastructure and applications.

In conclusion, this chapter provided valuable insights into how OPA and Harbor can be used together to ensure the security and policy compliance of modern DevOps workflows. In the next chapter, we will evaluate the application layer vulnerabilities and weaknesses within the context of cloud-native security. We will learn about secure coding practices and security components that should be used at the application level, thus avoiding any potential security breaches and attacks.

Quiz

- What does OPA provide policy enforcement of in an organization's infrastructure?

- What are some of the use cases for OPA from a compliance standpoint?

- What makes Harbor a valuable tool to ensure the security of applications in production?

- How can OPA and Harbor be used together to enhance the security and policy management of an infrastructure?

- What are the benefits of using Harbor in a DevOps workflow?

Further readings

You can refer to the following links for more information on the topics covered in this chapter:

- `https://mihirshah99.medium.com/stock-manipulation-by-exploiting-serverless-cloud-services-9cea2aa8c75e`

- `https://github.com/goharbor/harbor`

- `https://docs.docker.com/registry/insecure/#use-self-signed-certificates`

- `https://github.com/goharbor/harbor/wiki/Architecture-Overview-of-Harbor`

- `https://goharbor.io/docs/1.10/install-config/configure-yml-file/`

- `https://www.openpolicyagent.org/docs/latest/kubernetes-introduction/`

3

Cloud Native Application Security

This chapter aims to provide an in-depth understanding of the security considerations involved in cloud-native application development. As the shift toward cloud-based application development continues to grow, it is crucial for software engineers, architects, and security professionals to understand the security implications and best practices for building secure cloud-native applications.

In this chapter, you will explore the various security practices that should be integrated into the development process of cloud-native applications. You will learn about the benefits of following agile methodologies and how security considerations can be incorporated into the planning stage. Additionally, you will delve into the importance of integrating security into the development process by discussing the **Open Web Application Security Project (OWASP)** methodologies and tools.

By the end of this chapter, you will understand the key security considerations and best practices for building secure cloud-native applications. You will have a good understanding of the different tools and technologies available to help secure cloud-native applications, and you will be able to integrate security into your own cloud-native development processes. This chapter covers the following:

- Overview of cloud-native application development

- Integrating security into the development process

- Supplemental security components

Technical requirements

In this section, you need to install the following tools and guides:

- OWASP ASVS v4.0

- Helm v3.11

- Vault v0.23.0

- Kubernetes v1.27

Overview of cloud-native application development

Cloud-native application development is a modern approach to software development that leverages cloud technologies and microservices architectures. With the increasing demand for more agile and scalable applications, cloud-native application development has become an increasingly popular choice for organizations.

However, with this approach comes new security implications that must be considered. As applications are broken down into smaller, independently deployable components, the attack surface increases, and security becomes more complex.

In this section, we'll provide an overview of cloud-native application development, explore the security implications of this approach, and help you understand the basics of cloud-native app development and its security considerations. By the end of this section, you will have a comprehensive understanding of the fundamentals of cloud-native application development and its impact on security. Let's start by learning how a traditional app development architecture is different from cloud-native app development architecture.

Differences between traditional and cloud-native app development

In *Chapter 1*, we discussed the need for cloud-native apps in great depth, and understood the rise of cloud-native architecture. In this section, we will learn about the actual implementation and architectural differences that software engineers are usually mindful of when creating cloud-native apps. To all security engineers, this provides a good idea of how to approach an application when undertaking a pentesting assessment. The architectural differences between cloud-native and legacy apps can be visualized as follows:

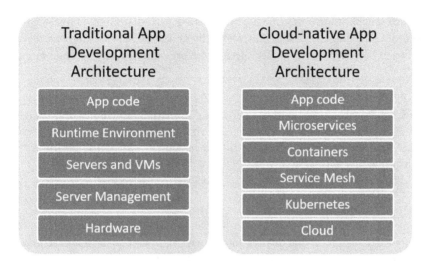

Figure 3.1 – Architectural differences between cloud-native and legacy apps

Traditionally, applications were developed as monolithic systems, with all components tightly coupled and running on a single machine. This approach was sufficient when applications were relatively simple, and the underlying hardware was dedicated to a single application. However, as applications have grown in complexity, traditional approaches have struggled to keep pace with the demands of modern software development.

Cloud-native application development, on the other hand, leverages cloud technologies and microservices architectures to build and deploy applications. In this approach, applications are broken down into smaller, independently deployable components, which can be developed, tested, and deployed independently. This allows for greater agility and scalability, as well as increased resilience in the face of failures.

One key difference between traditional and cloud-native application development is the use of containers. In cloud-native app development, containers provide a lightweight and portable way to package applications and their dependencies. This allows for greater flexibility in terms of deployment, as well as improved resource utilization compared to traditional virtualization approaches.

Another difference is the use of microservices architectures, which allow for the more granular scaling and management of individual components. In traditional applications, adding new functionality often requires changes to the entire application, while in cloud-native app development, new services can be added without affecting the rest of the application.

Overall, cloud-native application development represents a departure from traditional approaches and offers many benefits, including increased agility, scalability, and cost efficiency. However, it also requires a different approach to security, as the attack surface increases, and security becomes more complex.

> **Important note**
>
> The key difference is that cloud-native prioritizes automation, scalability, and resiliency. It utilizes the principles and tools of DevOps, containerization, and microservices to achieve these goals. This contrasts with traditional application development, which may prioritize stability and control over agility and scalability.

The DevOps model

DevOps is a software development and delivery model that emphasizes collaboration, communication, and integration between development and operations teams. This model aims to reduce the time to market and increase the efficiency of the software delivery process by integrating the development and operations functions.

In the DevOps model, development and operations teams work together to automate and optimize the software development process, from writing and testing code to deploying and managing it in production. This results in the faster delivery of high-quality software, improved collaboration and communication between teams, and the more efficient use of resources.

The DevOps model is often associated with cloud computing and cloud-native application development, as it enables organizations to take advantage of the scalability, agility, and cost-effectiveness of the cloud.

DevOps practices, such as **continuous integration/continuous delivery and deployment (CI/CD)**, allow organizations to automate and streamline their software delivery processes, while also improving the security and reliability of their applications. This results in improved customer satisfaction, faster time to market, and reduced costs. All of the steps in the DevOps **Software Development Life Cycle (SDLC)** process can be visualized in the following diagram:

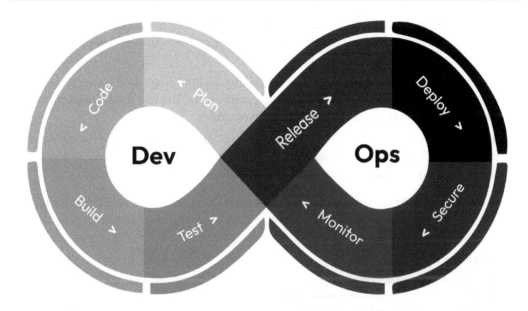

Figure 3.2 – Agile DevOps SDLC process

The DevOps model for software delivery is a continuous loop that encompasses the following stages:

1. **Plan**: In this stage, the requirements and objectives of the software are defined, and the scope and timeline of the project are determined. This is a crucial step for ensuring that the project stays on track and meets the desired outcomes.

2. **Code**: During this stage, the actual coding of the software takes place. Developers write the code that implements the functionalities outlined in the planning stage.

3. **Build**: In this stage, the code is compiled and transformed into a build that is ready to be tested. This stage involves integrating all the different components of the code and ensuring that everything works together seamlessly.

4. **Test**: This stage is where the build is subjected to various tests to verify its functionality and ensure that it meets the requirements. This stage is crucial for catching any bugs or issues that could impact the user experience.

5. **Release**: If the build passes all the tests, it is ready for release. The release stage involves making the software available for deployment to end users.

6. **Deploy**: During this stage, the software is deployed to production and made available to end users. This is a critical step that requires careful planning and coordination to ensure a smooth deployment process.

7. **Secure**: The security of the software is a top priority, and this stage involves implementing various security measures to protect the software and its users. This may include implementing encryption, access control, and other security technologies.

8. **Monitor**: The final stage of the DevOps model involves ongoing monitoring of the software to ensure that it continues to perform optimally. This stage involves collecting data, analyzing it, and making any necessary adjustments to the software to keep it running smoothly.

By following this loop, software development and delivery teams ensure that they are delivering high-quality software that meets the needs of the end users while also prioritizing security and performance.

> **Important note**
>
> The DevOps model is a continuous loop, meaning that it does not end after the deployment and securing stage. It is important to continuously monitor the application and its environment for any potential security risks, vulnerabilities, or incidents. The monitoring stage should feed back into the planning stage to adjust and make improvements to the overall process. This creates a culture of continuous improvement and supports the integration of security into the development process.

Cloud-native architecture and DevOps

A cloud-native architecture provides a robust foundation for DevOps practices by enabling teams to rapidly deploy and iterate applications at scale. The cloud-native approach leverages containerization, microservices, and CI/CD pipelines to streamline the software delivery process and minimize downtime.

In a cloud-native context, the DevOps loop is amplified by the ability to spin up new service instances quickly and easily, test and validate changes in a safe and isolated environment, and then roll out updates to production with minimal disruption. This empowers teams to quickly respond to changing business needs and customer requirements, and drive innovation at a faster pace.

The cloud-native architecture also provides robust security and monitoring capabilities, enabling teams to monitor the health and performance of their applications in real time and respond quickly to any potential threats. This, in turn, helps teams to maintain the integrity and security of their applications throughout the DevOps life cycle.

In short, the cloud-native architecture provides the ideal framework for DevOps practices, enabling teams to deliver applications faster, with greater reliability and security.

Introduction to application security

Application security is the practice of creating and maintaining the security of software applications. It encompasses a wide range of security controls, processes, and technologies, which are designed to prevent and mitigate various types of threats to an application's availability, confidentiality, and integrity.

Application security starts with the development of secure software, which requires following secure coding practices and incorporating security requirements into the design phase. It also involves securing the infrastructure that the application is deployed on and implementing security controls to protect against external threats such as network attacks and malicious code execution.

Some of the common application security measures include the encryption of sensitive data, input validation and sanitization, authentication and authorization, access control, and logging and monitoring.

It is important to note that application security is a continuous process that evolves as new threats emerge and technology changes. As such, it requires the regular testing, updating, and monitoring of security controls and practices.

In today's digital landscape, where applications play a critical role in business operations and user experience, organizations need to take application security seriously and invest in the necessary resources and processes to ensure their applications are secure.

Overview of different security threats and attacks

The cloud-native space has its own set of security threats and attacks that are unique to the architecture and environment of cloud-native applications. For a cloud-native system design, the factors could be visualized into three components:

- **Infrastructure security**: This category of security threats and attacks primarily targets the underlying infrastructure that supports cloud-native applications
- **Network security**: This category of security threats and attacks targets the network infrastructure that supports cloud-native applications
- **Application security**: This category of security threats and attacks targets cloud-native applications directly

However, since, in the cloud-native space, where development and operations are very tightly bound – meaning the development of the application, setting up the infrastructure, and cloud infrastructure management are performed in an automated fashion and at the code level – all three tiers of security fall under one single framework, and everyone working on the business application is equally responsible for all areas of security. Hence, this requires very good collaboration across teams to maintain the security of the application. The security components for building a robust cloud-native security solution can be visualized as follows:

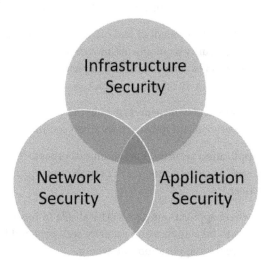

Figure 3.3 – Cloud security components

The overlapping region encompasses cloud-native security. We can further provide examples for each component of security as follows:

- **Application security**: Application security in the cloud-native space focuses on securing the application layer of the software stack, as opposed to the lower-level infrastructure components. This includes securing the code base and the application runtime environment, as well as protecting the data and resources that the application interacts with. In the cloud-native space, application security requires a deep understanding of the unique challenges posed by distributed, containerized applications, and must be integrated into the SDLC from the earliest stages of design and development. This includes practices such as threat modeling, security-focused coding and testing, and continuous monitoring and assessment to detect and remediate vulnerabilities as they arise. Effective application security in the cloud-native space requires close collaboration between developers, operations teams, and security professionals, as well as the use of specialized tools and technologies designed for the unique challenges of cloud-native application environments. The initial steps for performing application security assessments are these:

- **Application vulnerabilities**: Cloud-native applications can be vulnerable to various types of attacks such as injection attacks, **cross-site scripting** (**XSS**), and broken authentication and session management. To mitigate these risks, application developers can use security-focused coding practices such as input validation and sanitization, output encoding, and secure session management.

- **Vulnerability management**: Cloud-native applications are highly distributed and complex, which makes it important to have a comprehensive vulnerability management program. This can include practices such as continuous vulnerability scanning and assessment, penetration testing, and regular security audits. It is also essential to have a system for prioritizing and addressing vulnerabilities based on risk severity.

- **Threat modeling**: Threat modeling is an important process for identifying potential security risks and vulnerabilities in cloud-native applications. This involves identifying the various components and dependencies of the application, understanding potential attack vectors, and evaluating the impact of an attack. This process can help to identify and prioritize potential vulnerabilities and can inform the development of security controls and countermeasures.

To ensure that an application in the cloud-native space is secure, it's important to take a comprehensive approach that covers vulnerabilities at every stage of the SDLC. This involves adopting security-focused coding practices, performing vulnerability management and threat modeling, and regularly monitoring and assessing the system to detect and mitigate new security risks. By taking these steps, you can minimize the risk of security breaches and help to ensure that your application and its associated infrastructure remain secure over time.

- **Network security**: Network security in the cloud-native space focuses on protecting the communication channels between the various components of a distributed application. This includes securing the data in transit, as well as the underlying network infrastructure that enables this communication.

Some common network security threats and vulnerabilities in the cloud-native space include misconfigured cluster networking, unauthorized access to the network, and data breaches resulting from network-level attacks. To address these threats, it is essential to implement network security measures such as encryption, access controls, and network monitoring.

One important aspect of network security in the cloud-native space is ensuring the security of container networking. Containers are often used to deploy and scale applications in the cloud, and they require their own network interfaces and routing rules. It is essential to ensure that the container network is properly secured, and that communication between containers is restricted only to necessary ports and protocols.

In addition to container networking, network security in the cloud-native space also includes securing the underlying infrastructure such as load balancers, **virtual private clouds** (**VPCs**), and other network components. This can include implementing security groups, firewalls, and intrusion detection systems to prevent unauthorized access and network-level attacks.

Overall, network security is a crucial aspect of cloud-native security and must be considered alongside infrastructure and application security to ensure a comprehensive and effective security posture.

The network security examples can further be broken down as follows:

- **Misconfigured cluster networking**: One of the biggest network security vulnerabilities in the cloud-native space is misconfigured cluster networking. This can result in unauthorized access to sensitive data and resources, as well as allow attackers to move laterally within a network. To mitigate this risk, it's important to implement network segmentation, limit network exposure, and ensure that security policies are correctly configured.

- **Cluster and node security**: Cloud-native infrastructure often involves large-scale clusters of nodes, which can be vulnerable to attacks such as node compromise, denial of service, and unauthorized access. To protect against these risks, it's important to implement security controls such as access control, authentication and authorization mechanisms, and encryption. It's also important to regularly monitor the health and security of nodes and clusters and quickly address any security incidents.

- **Monitoring and logging**: Effective monitoring and logging are critical for network security in the cloud-native space. This involves monitoring the network for potential threats and attacks, as well as logging all network activity for later analysis and forensic investigation. This can help to identify and respond to security incidents quickly and can also provide valuable data for ongoing risk assessment and vulnerability management. It's important to ensure that logging and monitoring tools are correctly configured and integrated into the overall security architecture.

- **Infrastructure security**: Infrastructure security in the cloud-native space involves protecting the underlying physical and virtual resources that support an application or service. This includes the security of the cloud service provider infrastructure, such as the security of servers, storage, and networking resources. Infrastructure security also involves securing the Kubernetes cluster and nodes, which form the foundation of a cloud-native application.

One of the key differences between infrastructure security and network security is the focus of security measures. Network security is focused on protecting data in transit across a network, while infrastructure security is focused on the security of the physical and virtual infrastructure that supports an application or service.

Infrastructure security in the cloud-native space requires a range of security measures such as identity and access management (IAM), secure configuration management, and secure network design. It is important to ensure that only authorized users have access to infrastructure resources, that the resources are configured securely, and that network design is optimized for security.

Effective infrastructure security in the cloud-native space requires a proactive approach, which involves the ongoing monitoring and assessment of security risks, as well as implementing and testing security controls and countermeasures to address identified vulnerabilities. A few of the examples involved in infra security in the cloud space are as follows:

- **IAM**: IAM is an important aspect of infrastructure security in the cloud-native space. IAM systems are used to manage user authentication, authorization, and access control for cloud resources. It involves defining roles and permissions for different users or groups, implementing **multi-factor authentication** (**MFA**), and enforcing security policies such as password strength requirements and session timeouts.

- **Container security**: Containers are a popular technology for deploying applications in the cloud-native space, but they can also introduce new security risks. Container security involves securing the container image, runtime environment, and host system. It can include practices such as scanning container images for vulnerabilities, restricting access to the container host, and implementing network segmentation and isolation to prevent lateral movement.

- **Infrastructure as code (IaC) security**: IaC is the practice of defining and managing infrastructure resources using code. IaC tools such as Terraform and CloudFormation are commonly used in the cloud-native space to automate the creation and configuration of cloud resources. However, just like any other code, IaC code can contain security vulnerabilities. IaC security involves applying security best practices to the code, such as implementing secure coding standards, using secure authentication and authorization mechanisms, and scanning for vulnerabilities in the code. It also involves enforcing security policies and access controls for the IaC tools and the cloud resources they manage.

Important note

The cloud-native space is constantly evolving, and new threats and vulnerabilities can emerge at any time. Therefore, it is important to maintain a proactive approach to security by continually monitoring, assessing, and updating security measures to ensure that they remain effective and relevant. In addition, security should be viewed as a shared responsibility, with all stakeholders involved in the development, deployment, and management of cloud-native applications and infrastructure, playing a role in ensuring security best practices are followed.

The following diagram provides a visual representation of all three components that comprise cloud-native security:

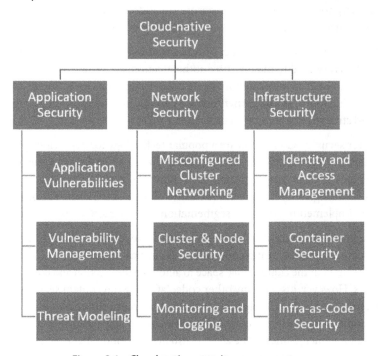

Figure 3.4 – Cloud-native security components

On realizing the different components that build up to delivering a cloud-native security solution for a product, let us understand how each component fits individually within the SDLC.

Integrating security into the development process

Integrating security into the development process is critical for ensuring that cloud-native applications are developed with security in mind from the outset. This approach, often referred to as *"shift-left"* security, involves incorporating security considerations and practices throughout the entire development life cycle, from design and development to deployment and maintenance.

By integrating security early in the development process, developers can identify potential security risks and vulnerabilities and proactively address them, rather than trying to fix them after the application is already deployed. This can save time and resources, as well as reduce the risk of security breaches and data loss.

Integrating security into the development process can also help to foster a culture of security awareness and responsibility within development teams, promoting a shared understanding of security risks and best practices. It can also help to reduce the burden on security teams by empowering developers to take ownership of security in their own code and applications.

Overall, integrating security into the development process is critical for developing secure, reliable, and resilient cloud-native applications, and is an essential part of any effective application security strategy. It is also important to understand how application security plays out in the cloud-native realm. OWASP has documented guidelines on understanding the top 10 security lapses leading to cloud-native environment compromise.

OWASP Top 10 for cloud native

OWASP Top 10 is a list of the most critical web application security risks, compiled by OWASP, a non-profit organization focused on improving the security of software. The OWASP Top 10 list serves as a guide for developers, architects, and security professionals to ensure that web applications are designed and developed with security in mind.

In the context of cloud-native security, the OWASP Top 10 can be applied to the design and development of cloud-native applications. The risks on the OWASP Top 10 list can manifest themselves in cloud-native environments in a variety of ways, such as through misconfigured Kubernetes clusters or vulnerabilities in container images.

For example, injection attacks, which are the number one risk on the OWASP Top 10 list, can occur in cloud-native environments if containers are not properly secured or API calls are not properly validated. Similarly, XSS and broken authentication and session management vulnerabilities can be introduced if cloud-native applications are not developed with security best practices in mind.

To ensure that cloud-native applications are secure, developers should be familiar with the OWASP Top 10 and take steps to mitigate these risks throughout the SDLC. This includes performing code reviews, using secure coding practices, and integrating security testing into the CI/CD pipeline. Additionally, security professionals should use the OWASP Top 10 as a guide for assessing the security posture of cloud-native environments and designing security controls and countermeasures to mitigate these risks.

Let's look at each of the OWASP Top 10 criteria individually and understand how they apply to the cloud-native space, while also trying to understand how an attacker could practically exploit a benign web app vulnerability to compromise the wider cloud infrastructure. If we consider OWASP Top 10 2021, the following vulnerabilities apply in the context of cloud security:

- **Broken access control**: Broken access control is a security vulnerability that occurs when an application doesn't properly enforce access controls, allowing attackers to access restricted resources or perform actions that they shouldn't be able to. A common example of this vulnerability is when an application uses URLs to control access to certain functions or resources. If the URLs are not properly secured, an attacker can manipulate them to access restricted areas or perform actions they shouldn't be able to.

For example, consider an e-commerce website that has different pages for different types of users – one for customers, one for employees, and one for administrators. If the website doesn't properly enforce access control, an attacker could use a URL manipulation technique to gain access to the administrative page without proper authentication. This could allow the attacker to perform actions that are only meant to be performed by administrators, such as modifying user accounts, accessing sensitive data, or changing site configurations.

In the context of cloud-native applications, broken access control vulnerabilities can be even more dangerous because of the distributed and dynamic nature of these applications. An attacker who gains access to one part of the system can potentially compromise the entire cloud-native landscape, especially if the access control between different parts of the system is not properly enforced. This could allow the attacker to perform remote code execution, steal sensitive data, or cause other types of damage.

To detect and prevent broken access control vulnerabilities, security measures such as **role-based access control** (**RBAC**), **attribute-based access control** (**ABAC**), and the principle of least privilege should be implemented. Cloud-native security tools such as **Open Policy Agent** (**OPA**), Kyverno, and Kubernetes RBAC can help enforce proper access controls in a cloud-native environment. Additionally, regular security testing and auditing can help identify and remediate any access control issues before they can be exploited by attackers.

- **Cryptographic failures**: Insecure cryptographic storage is a vulnerability that occurs when sensitive data is not encrypted or encrypted with weak algorithms, which can lead to unauthorized access or the manipulation of data. A real-world scenario of this vulnerability could be an online shopping website that stores users' credit card information in plaintext format on their servers. If an attacker gains access to the server, they can easily read and steal credit card information, which can then be used for fraudulent purchases.

  In the context of cloud-native applications, this vulnerability can be exacerbated due to the dynamic and distributed nature of the environment. Cloud-native applications often rely on microservices and APIs that communicate with each other over the network. If encryption and decryption keys are not managed properly or the encryption algorithms are weak, attackers can intercept and manipulate sensitive data as it moves between services.

  Attackers can carry out remote code execution by exploiting this vulnerability and injecting malicious code into the application that allows them to read and write data that should be protected. They can also use this vulnerability to compromise the entire cloud-native landscape by accessing sensitive data in other parts of the system, such as credentials, keys, and certificates.

  To avoid this vulnerability, it is important to use strong encryption algorithms and manage encryption and decryption keys properly. In addition, sensitive data should be stored separately from other data and should be accessible only to authorized users. Cloud-native tools that can help prevent this vulnerability include the following:

- **Key management services** such as AWS KMS or HashiCorp Vault, which allow for secure key storage and management

- **Encryption libraries** such as OpenSSL or Bouncy Castle, which provide strong encryption algorithms and protocols

- **Security testing tools** such as OWASP **Zed Attack Proxy** (**ZAP**) or Burp Suite, which can help identify insecure cryptographic storage and other vulnerabilities in the application code

- **Injection**: Injection vulnerability is a type of security vulnerability that allows an attacker to inject malicious code or commands into an application, which can then be executed by the application's backend server. The attacker can use this vulnerability to gain unauthorized access to sensitive data or execute malicious code that can compromise the entire system.

  A real-world example of injection vulnerability is SQL injection, which is a common attack vector for web applications. In this scenario, an attacker can inject SQL commands into a web application's input field (such as a search bar or a login form) to gain access to the application's backend database. For example, an attacker could insert a SQL command that instructs the database to retrieve all records from a table, including sensitive data such as usernames and passwords.

  In the context of cloud-native applications, injection vulnerabilities can be even more dangerous because they can potentially allow attackers to carry out remote code execution and compromise the entire cloud-native landscape. For example, an attacker can inject malicious code into a cloud-native application's code repository, which can then be automatically deployed to multiple containers or nodes in a cluster. If the code contains a vulnerability that allows remote code execution, the attacker can potentially compromise the entire cloud-native landscape.

  To detect and prevent injection vulnerabilities in cloud-native applications, it is important to use security-focused coding practices such as input validation and sanitization, output encoding, and secure session management. Additionally, tools such as OWASP's Dependency-Track and CNCF's Falco can be used to scan application dependencies and detect potential vulnerabilities in real time.

  To mitigate the risk of injection vulnerabilities, cloud-native applications can use security-focused coding practices, perform regular vulnerability scans and assessments, implement access controls and authentication mechanisms, and follow industry-standard security frameworks such as OWASP or CNCF. Additionally, cloud-native tools such as CNCF's Notary and OPA can be used to enforce secure deployment and policy controls to minimize the attack surface.

- **Insecure design**: Insecure design is a broad category that encompasses vulnerabilities stemming from poor architecture or design choices in the application. A common example of an insecure design vulnerability is the lack of proper input validation, which allows malicious actors to insert code into input fields to exploit application vulnerabilities.

An example of a real-world scenario where insecure design resulted in a major breach was the Equifax breach of 2017, where attackers exploited a vulnerability in an Apache Struts web framework to gain access to the sensitive data of more than 143 million people.

In cloud-native applications, insecure design can manifest in various ways, such as the use of unsecured APIs, inadequate access control mechanisms, and insecure data storage practices. Attackers can exploit these vulnerabilities to gain access to sensitive data, execute code remotely, or compromise the entire cloud-native infrastructure.

To prevent insecure design vulnerabilities, it is essential to ensure that security is a fundamental consideration in the application design and development process. Cloud-native tools such as Kubernetes can help enforce security best practices by providing secure defaults for containers and other resources, as well as providing features such as RBAC and Pod security policies to limit access to resources and reduce the attack surface.

Furthermore, it is essential to conduct regular security assessments and penetration testing to identify and remediate insecure design vulnerabilities proactively. By regularly testing and fixing design vulnerabilities, you can strengthen your application security posture and reduce the risk of a security breach.

Some of the secure design principles that can help to prevent insecure design vulnerabilities are as follows:

- **Principle of least privilege**: Users should only be given the minimum level of access necessary to perform their duties

- **Defense in depth**: This principle advocates for multiple layers of security controls, which ensures that if one layer fails, there are still other layers to fall back on

- **Separation of concerns**: The system should be designed so that each component has a single, well-defined responsibility

- **Fail securely**: The system should be designed so that it fails in a secure manner, instead of failing open and allowing attackers to exploit the system

- **Secure defaults**: All the settings of the system should be secure by default

- **Secure communication**: Communication between different components of the system should be secure

Now, let's apply these principles to the insecure design vulnerability scenario. Insecure design vulnerabilities arise when there are flaws in the design of the application, leading to potential security gaps. For example, if an application does not properly validate user input, it could lead to SQL injection attacks.

To avoid this vulnerability, we need to ensure that the application is designed with security in mind from the beginning of the development process. This includes ensuring that each component has a single, well-defined responsibility, with a minimum level of access necessary to perform its duties. We should also ensure that secure defaults are used throughout the system and that all communication is secure.

In terms of specific tools and techniques, we can use secure design methodologies such as threat modeling, penetration testing, and code reviews to identify potential vulnerabilities. We can also use cloud-native tools such as Kubernetes, which provides RBAC, and OPA, which can help enforce secure design principles.

Overall, by incorporating secure design principles and using cloud-native security tools, we can prevent insecure design vulnerabilities in our cloud-native applications.

- **Security misconfiguration**: Security misconfiguration refers to the improper configuration of an application or its components, which can leave them vulnerable to exploitation by attackers. A real-world scenario for this vulnerability would be a web application that has a default username and password for a database or web server. If the default credentials are not changed, an attacker can easily gain access to the server or database, potentially leading to a data breach.

  In the context of cloud-native applications, security misconfiguration can be even more damaging, as the dynamic nature of cloud environments can create additional attack surfaces.

  For example, misconfigured container images or cloud services can lead to unauthorized access to data or resources. Additionally, the use of IaC tools can introduce security misconfigurations if not properly configured.

  Attackers can exploit security misconfigurations to gain remote code execution by taking advantage of misconfigured cloud services or by injecting malicious code into vulnerable components. Once an attacker has gained access to a cloud-native application, they can potentially compromise the entire cloud-native landscape by moving laterally through the environment and exploiting other vulnerabilities.

  To detect security misconfigurations, organizations should implement automated vulnerability scanning and perform regular security audits. Measures to prevent security misconfigurations include following security best practices for cloud services, maintaining up-to-date software versions, and implementing secure default settings for applications.

  Several cloud-native tools from the CNCF and OWASP can be used to avoid security misconfigurations, including the following:

  - **Kubernetes admission controllers**: This is a feature in Kubernetes that allows administrators to define custom validation rules for incoming requests to the Kubernetes API server. This can help prevent security misconfigurations in the cluster.

- **OPA**: This is a policy engine for cloud-native environments that can be used to enforce security policies across the infrastructure. OPA can help prevent security misconfigurations by ensuring that infrastructure resources are properly configured and secured.

- **Terraform**: This is an IaC tool that can be used to define and manage cloud infrastructure. By defining infrastructure resources as code, Terraform can help prevent misconfigurations and ensure that resources are properly secured.

- **Docker Bench Security**: This is a tool from Docker that can be used to audit Docker containers and images for security vulnerabilities and misconfigurations.

- **OWASP ZAP**: This is a popular web application security scanner that can be used to identify security misconfigurations in web applications.

These tools can help detect and prevent security misconfigurations in cloud-native environments and can help ensure that infrastructure resources are properly configured and secured.

- **Vulnerable and outdated components**: Vulnerable and outdated components are a common vulnerability in software development, where an application is built using third-party libraries or components that have known security vulnerabilities. Attackers can exploit these vulnerabilities to compromise the application and gain unauthorized access to sensitive data or perform other malicious actions.

  Software Composition Analysis (SCA) tools can be used to detect and manage outdated or vulnerable components in software applications. These tools are designed to scan the application code and its dependencies for known vulnerabilities, outdated libraries, and other potential security issues.

  Using **SCA** tools can help identify vulnerable and outdated components early in the development cycle, enabling teams to take proactive measures to mitigate the risks. SCA tools can also be integrated into the DevOps pipeline to provide continuous monitoring and reporting of the software components used in the application.

  For example, a real-world case of vulnerable and outdated components vulnerability is the Equifax data breach in 2017. In this case, the attackers were able to exploit a vulnerability in Apache Struts, a popular open source web application framework that was used in Equifax's web application. The vulnerability was patched in March 2017, but Equifax failed to apply the patch, leaving their system exposed to the attack.

  In a cloud-native application, the use of container images and microservices architecture can introduce additional challenges in managing and securing the software components. It is important to implement a process for regularly scanning and updating the container images and microservices used in the application.

Cloud-native tools such as Kubernetes can be used to manage container images and ensure that the latest security patches are applied. SCA tools such as Snyk, Sonatype, and Black Duck can be integrated into the pipeline to scan the container images and identify any outdated or vulnerable components.

In addition to using SCA tools, it is important to establish policies and processes for managing software components. This includes creating an inventory of all the components used in the application, establishing guidelines for selecting and using third-party components, and regularly reviewing and updating the components to ensure that they are secure and up to date. By adopting these practices and using SCA tools, organizations can reduce the risk of vulnerabilities resulting from outdated or vulnerable components.

- **Identification and authentication failures**: Broken identification and authentication vulnerability refers to security weaknesses in the way user authentication and session management are handled in an application. A real-world scenario for this vulnerability could be an e-commerce website that fails to invalidate a user's session after they log out, allowing an attacker to gain access to the user's account after the user has logged out.

  In a cloud-native application, attackers can exploit this vulnerability by stealing a user's session tokens or credentials and using them to impersonate the user. This could lead to an attacker carrying out remote code execution and potentially compromising the entire cloud-native landscape.

  To detect and prevent broken authentication and session management vulnerabilities, applications should implement strong password policies, use MFA, and regularly rotate session tokens. Other measures include setting short expiration times for session tokens, enforcing proper logout procedures, and implementing proper error handling for failed authentication attempts.

  Several cloud-native tools from the CNCF and OWASP can be used to prevent broken authentication and session management vulnerabilities. For example, Keycloak is an open source IAM solution that can be used to authenticate users and manage sessions securely. Other tools that can be used to mitigate this vulnerability include OWASP ZAP, which can be used to scan for authentication and session management vulnerabilities in web applications, and Istio, which can be used to enforce access control policies at the network level.

- **Software and data integrity failures**: Software and data integrity failures refer to the failure to properly protect the integrity of software and data. This can occur when systems are improperly configured or updated, or when malicious actors intentionally introduce unauthorized changes to software or data. As a result, it can lead to devastating consequences, such as data loss, financial loss, and reputational damage.

  A real-world example of this vulnerability is the 2017 NotPetya attack, which targeted a Ukrainian software firm that provided tax and accounting software to the Ukrainian government. Attackers used a backdoor in the company's software update system to distribute malware, which spread rapidly through networks, causing widespread disruption and financial damage to organizations around the world.

Another real-world example would be the infamous data breach of Deliveroo. In 2017, the food delivery service company Deliveroo suffered a data breach that resulted in the compromise of thousands of customers' personal information. The attackers were able to gain unauthorized access to one of Deliveroo's **Amazon Web Services** (**AWS**) accounts and used the access to steal customers' information such as email addresses, phone numbers, and partial payment card details.

The attackers exploited a vulnerability in Deliveroo's website to gain access to an AWS key, which was then used to access the AWS account. The vulnerability was caused by a flaw in the website's code, which did not properly validate user input, allowing the attackers to inject malicious code into the site and gain access to the key.

This breach was a result of a failure in both software and data integrity. Deliveroo's website contained a vulnerability that allowed the attackers to gain access to the AWS key, and once they had access, they were able to exfiltrate sensitive customer data. This highlights the importance of ensuring that software and data are kept secure and that proper security measures are in place to prevent unauthorized access.

To prevent such attacks, it is important to follow secure coding practices, such as input validation and output encoding, and regularly update and patch software and systems to prevent known vulnerabilities. Additionally, implementing strong access control and monitoring tools can help detect and prevent unauthorized access to sensitive data.

There are also tools available to help organizations detect and prevent software and data integrity failures, such as static code analysis tools, **dynamic application security testing** (**DAST**) tools, and **runtime application self-protection** (**RASP**) tools. The **Cloud Security Alliance** (**CSA**) and OWASP also provide resources and guidelines for securing cloud-native applications and preventing data breaches.

In the context of cloud-native applications, software and data integrity failures allow attackers to gain remote code execution and compromise the entire cloud-native landscape. For example, if an attacker can modify a container image or the code running inside a container, they may be able to gain access to sensitive data, escalate privileges, or launch further attacks.

To detect and prevent software and data integrity failures, it's important to implement strong security measures throughout the entire SDLC, including secure coding practices, vulnerability scanning, and continuous monitoring of systems for unauthorized changes. Cloud-native tools such as Kubernetes admission controllers, Falco, and Anchore can help detect and prevent unauthorized changes and enforce policy-based security controls.

Some measures that can be taken to avoid these attacks include performing regular software and system updates, minimizing the use of untrusted software components, using secure coding practices and techniques, implementing strong access control, and implementing a robust backup and disaster recovery plan. Additionally, organizations can use SCA tools such as WhiteSource, Snyk, and Sonatype to detect and remediate vulnerable software components before they are deployed to production.

- **Security logging and monitoring failures**: Security logging and monitoring failures is an OWASP Top 10 vulnerability that relates to inadequate logging, monitoring, or analysis of security events in the system, leading to a lack of visibility into system activity and a reduced ability to detect and respond to security incidents.

  A real-world example of this vulnerability was the 2013 Target data breach. In this case, Target had a security monitoring system in place that was designed to detect and alert on suspicious network activity. However, the system was not properly configured and monitored, so when attackers gained access to Target's network and began exfiltrating sensitive customer data, the security team failed to respond to the alerts generated by the system.

  In the context of cloud-native applications, security logging and monitoring failures can be particularly dangerous. Because cloud-native environments are highly distributed and dynamic, security teams may lack visibility into the activity of individual containers, microservices, or functions. Additionally, the use of container orchestrators such as Kubernetes can obscure the underlying infrastructure, making it difficult to identify which resources are hosting which applications or services.

  Attackers can exploit security logging and monitoring failures to gain remote code execution and compromise the entire cloud-native landscape. For example, an attacker could use a vulnerability in an application to gain access to a container and then use that container to move laterally across the cluster and compromise other containers. Without proper logging and monitoring, security teams may not be able to detect these actions until it is too late.

  To avoid security logging and monitoring failures, it is important to implement logging and monitoring best practices. This includes ensuring that logs are generated and stored securely, that all relevant events are logged, and that the logs are regularly reviewed and analyzed for signs of suspicious activity.

  Several cloud-native tools can be used to mitigate this vulnerability, including the following:

  - **Prometheus**: A monitoring system and time series database designed specifically for cloud-native environments
  - **Grafana**: A dashboard and visualization tool that integrates with Prometheus to provide real-time visibility into the health and performance of cloud-native applications
  - **Fluentd**: A data collection tool that allows logs to be collected from multiple sources, unified, and forwarded to other tools or storage systems
  - **Elastic Stack**: A collection of tools including Elasticsearch, Kibana, and Logstash that can be used to collect, store, and analyze logs in a cloud-native environment

In addition to these tools, it is important to ensure that all applications and services are designed with security in mind and that security logging and monitoring are integrated into the development process from the outset. By taking a proactive approach to security, organizations can minimize the risk of security logging and monitoring failures and ensure that they are well prepared to detect and respond to security incidents.

- **Server-Side Request Forgery** (**SSRF**): SSRF is a type of vulnerability that allows attackers to make unauthorized requests from a server by manipulating the server-side application to send requests to other servers. This can lead to data exposure and information theft, and can be used as a stepping stone for further attacks. In 2019, Capital One, a major financial institution, suffered a major data breach that was the result of an SSRF vulnerability.

In the Capital One breach, a former AWS employee was able to gain access to Capital One's customer data through an SSRF vulnerability. The attacker used a tool to scan for open ports on the Capital One server, identified the AWS metadata service, and then used the SSRF vulnerability to steal sensitive customer information from the AWS S3 storage bucket. The attacker was able to exfiltrate the personal data of over 100 million customers, including names, addresses, credit scores, social security numbers, and more.

Cloud-native applications are particularly vulnerable to SSRF attacks because of the dynamic and distributed nature of cloud-native architectures. In a cloud-native environment, applications are made up of multiple microservices and run on a cluster of servers, which are connected through various APIs and services. As a result, there are more attack surfaces and more opportunities for SSRF vulnerabilities to be introduced.

Attackers can gain remote code execution and compromise the entire cloud-native landscape through SSRF vulnerabilities by using the access granted by the initial request to send additional malicious requests to other servers within the network. Once they have control of a network, they can use it as a platform for more complex attacks, such as privilege escalation, data theft, or denial of service attacks.

To detect and prevent SSRF attacks, several measures can be taken. These include properly configuring firewalls, using input validation to ensure only legitimate requests are being sent, and implementing secure coding practices that take SSRF vulnerabilities into account. Tools such as the OWASP ZAP and **Cloud-Native Application Bundles** (**CNABs**) can also be used to automate the detection and remediation of SSRF vulnerabilities.

In the case of cloud-native applications, it is important to have a clear understanding of the interdependencies between services and to use tools such as AWS **Security Token Service** (**STS**) and RBAC to ensure that only authorized services are communicating with each other. Additionally, organizations should implement an IAM strategy that uses MFA and access policies to limit the exposure of sensitive data and resources.

In summary, SSRF is a critical vulnerability that can have severe consequences for organizations, particularly those that are running cloud-native applications. By taking proactive measures to detect and prevent SSRF vulnerabilities, organizations can reduce the risk of attacks and protect their customers' sensitive data.

Now that we've gained a very intricate understanding of the OWASP Top 10 for cloud-native, we can explore the tools available at our disposal that should be used and made part of the development process and the DevOps pipeline moving forward.

Not shift-left

In the traditional waterfall development process, security audits were often done only during deployment or production, which meant that security issues were expensive and time-consuming to fix. To address this problem, the concept of shifting left was introduced, which involves moving security controls earlier in the development process so that issues can be found and remediated sooner, resulting in lower costs.

However, in the DevOps era, the concept of shifting left is not as straightforward. DevOps embraces a continuous process that doesn't have a clear *"left"* or *"right"* and accepts that some bugs will only be found in production. The following diagram represents the traditional SDLC and how it contrasts with the modern DevOps era; this approach is focused on faster delivery cycles, and it relies on methodologies such as observability to help find issues post-deployment:

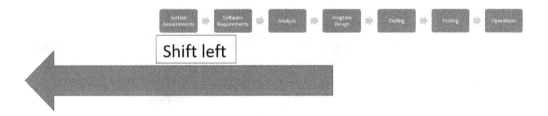

Figure 3.5 – Software development waterfall model

Moreover, the shift-left concept doesn't reflect the change in ownership and drive for independent teams that DevOps brings. The important change is not just shifting technical testing to the left but shifting the ownership of testing to the development team. This means that each development team should be equipped and empowered to decide the best place and time to run security tests, adapting it to their workflows and skills.

Therefore, instead of solely focusing on shifting left, it is essential to adopt an empowering security practice that enables the development team to take ownership of security testing. This means that security teams should provide the necessary tools and training to empower the development teams to run security tests on their code and provide guidance on how to address vulnerabilities that are found.

Overall, the concept of shifting left is still valid and essential, but it needs to be adapted to the continuous nature of DevOps and the shift toward independent teams. By empowering development teams to take ownership of security testing, organizations can enable faster delivery cycles while maintaining security and reducing the cost and time required to fix security issues.

All tooling and processes developed for security improvements should be in the context of a developer.

Security and development trade-off

Security is often perceived as a team in opposition to development practices; however, that ought not be the case. While we will discuss this in depth in *Chapter 4*, it is often imperative for security engineers to ask the following questions:

- What are we trying to secure and why is it essential that we secure it?
- Do we completely understand the working of the system/software to make a call?
- How would we implement a security solution, and is it scalable?
- Will there be a trade-off with the efficiency of the system/software?
- What could be the downsides of adding this security layer?
- Is the security-demanded engineering time worth the cost to implement the security solution?

As budding security engineers, we have the desire to secure and fix everything, even something that doesn't really need security attention. One of the ways to circumvent this is to wear the developer's hat and gain some context behind the software. Please note that you don't need to be an expert; just understand the working of the code, and then put yourself in the shoes of the developers to understand whether what was perceived as a security solution really is a security solution. One simple way to know the answer to that is to ask a question – can you or someone else exploit that weakness? If the answer to that is yes, then it might be worth having the wider team investigate it.

One of the ways to enumerate and identify security threats and weaknesses is by using automated tooling. While it may have a higher volume of false positives and false negatives, it does help security engineers to confine their landscape to the tool's findings. We won't go deeper into the trade-offs and understanding of the logical segment of using different tools. Instead, we are going to look into one such tool used for application security known as OWASP **Application Security Verification Standard (ASVS)** in the next section.

Supplemental security components

Multiple layers and other security components comprise part of a complete application security program. It is crucial that a skilled AppSec engineer is aware of all of these security protocols and systems in place. We are going to explore a few of them in this section within the context of securing the cloud-native space, starting with OWASP's flagship security guidelines for application security – ASVS.

OWASP ASVS

OWASP ASVS is a community-driven open source framework designed to help organizations assess the security of their web applications. The standard provides a checklist of security requirements that web applications should meet to ensure their security posture.

The framework is structured around three levels of increasing security coverage, with each level adding additional security controls. The levels are based on the sensitivity of the application and the data it handles, as well as the potential impact of a successful attack.

To use ASVS, an organization can compare its application to the standard's requirements and determine whether the necessary controls are in place to meet the security requirements. The framework can be used to assess both new and existing applications and can help organizations identify security gaps that need to be addressed.

When it comes to cloud-native applications, the ASVS framework can be a valuable tool for assessing the security of microservices, APIs, and other components in a cloud environment. By assessing the security requirements outlined in ASVS, organizations can ensure that their cloud-native applications are secure and that the components are properly configured to meet the necessary security controls.

Moreover, as cloud-native applications are typically built using microservices and other cloud technologies, using ASVS can help ensure that each microservice meets the required security controls. Additionally, by ensuring that each microservice is secure, it can reduce the risk of vulnerabilities being exploited in other components of the application.

OWASP ASVS is not a tool but rather a set of guidelines and requirements for testing the security of web applications. However, there are various tools available that can help implement the OWASP ASVS requirements.

To run OWASP ASVS tests, you can use a combination of automated and manual testing techniques. Here are the general steps to implement and run the OWASP ASVS tests:

1. *Familiarize yourself with the ASVS requirements*: The first step in using ASVS is to familiarize yourself with the requirements for each level. This will help you understand what tests are needed and how they should be performed.

2. *Identify the scope of the application*: Determine which parts of the application are in scope for testing, based on the requirements of the project.

3.   *Choose an appropriate testing methodology*: Determine the appropriate testing methodology to use, such as black-box, white-box, or gray-box testing.

4.   *Use automated tools*: Automated testing tools can be used to cover a wide range of OWASP ASVS requirements. Some of the tools that can be used include Burp Suite, OWASP ZAP, and Nmap.

5.   *Perform manual testing*: Manual testing is required for some of the requirements. This includes testing for business logic flaws, authentication, authorization testing, and more.

6.   *Document findings*: Document the results of the tests performed and classify any issues found according to the severity level.

7.   *Remediate findings*: Work with the development team to remediate any issues found during testing.

Regarding how the findings detected from the tool relate to the cloud-native space, OWASP ASVS is technology-agnostic and does not focus on any specific platform or deployment model. However, the guidelines can be applied to cloud-native applications as well since the same security principles apply regardless of the deployment model. The use of automated testing tools is especially useful in the cloud-native space, where applications can be scaled quickly and dynamically, making it difficult to maintain the security posture manually.

Some examples of findings that can be identified by OWASP ASVS are as follows:

- **Authentication**: The verification of the authentication mechanism, password complexity, and enforcement of password policies

- **Access control**: The verification of user role separation and permissions and the enforcement of access controls based on user roles

- **Input validation**: The verification of input validation and sanitization techniques to avoid common web application vulnerabilities such as SQL injection and XSS

- **Cryptography**: The verification of encryption key management, the usage of secure cryptographic algorithms, and the implementation of secure password storage

- **Error handling and logging**: The verification of error messages and logging to ensure that sensitive data is not exposed and that logs are secure from unauthorized access

- **Session management**: The verification of the mechanisms for session management such as timeouts and secure session token generation and handling

- **Data protection**: The verification of data protection mechanisms such as data classification, data retention policies, and secure deletion of data

- **API security**: The verification of the security of APIs such as the enforcement of API authentication and access control, input validation, error handling and logging, and rate limiting

- **Deployment and configuration management**: The verification of secure deployments and the configuration of servers and services, secure network communication, and the management of security incidents

These findings are just a small sample of the types of security issues that can be detected by OWASP ASVS. By identifying and addressing these issues, organizations can better secure their web applications and protect sensitive data.

Secrets management

Secrets management is an important security concern in cloud-native environments. Secrets are any sensitive data that needs to be kept confidential, such as passwords, tokens, keys, and certificates. In cloud-native environments, secrets can be used to authenticate and authorize access to various resources and services, such as databases, APIs, and cloud storage.

Proper secrets management is essential to ensure that this confidential data is protected from unauthorized access and use. One key aspect of secrets management is secure storage, which involves storing secrets in an encrypted format, both in transit and at rest. This is typically done using a secret store or a secret management service, such as HashiCorp Vault or Kubernetes Secrets, which can securely store and manage secrets at scale.

Another key aspect of secrets management is secure distribution, which involves securely transmitting secrets to the intended recipient. This can be challenging in cloud-native environments, where applications and services can be dynamically deployed and scaled up and down in response to changing demands. As a result, secrets distribution must be automated and integrated into the deployment and orchestration process.

In addition, secrets rotation is also an important consideration in cloud-native environments. This involves changing the secrets regularly to minimize the risk of unauthorized access. Automated secrets rotation can be implemented using tools such as Kubernetes Secrets or HashiCorp Vault.

Overall, secrets management is a critical security concern in cloud-native environments. By ensuring that secrets are stored, distributed, and rotated securely, organizations can protect their applications and services from unauthorized access and potential security breaches.

We will explore HashiCorp Vault and Kubernetes for secrets storage and management.

HashiCorp Vault

HashiCorp Vault is a popular tool used for managing secrets in cloud-native applications. Here are some common use cases for implementing Vault:

- **Dynamic secrets**: Vault allows dynamic secrets to be generated for external systems (such as databases or cloud services) on demand, instead of relying on long-lived static secrets

- **Encryption and decryption**: Vault can be used to encrypt and decrypt data, making it easier to store sensitive data in configuration files or other areas where it would otherwise be vulnerable

- **Access control**: Vault provides a comprehensive access control system, allowing administrators to restrict access to secrets based on policies and roles

- **Secret rotation**: Vault provides automated secret rotation capabilities, which can be used to regularly update secrets, reducing the risk of a compromised secret being used to access sensitive data

- **Secure storage**: Vault provides secure storage for sensitive data, with encryption at rest and other security features designed to protect against data leaks and other threats

Installing HashiCorp Vault on Kubernetes

To install HashiCorp Vault on a Kubernetes cluster, you can follow these general steps:

1. Deploy a Kubernetes cluster if you haven't already done so. You can use a managed Kubernetes service from a cloud provider or set up your own cluster using a tool such as `kops`, `kubeadm`, or `minikube`. We will use Docker Desktop for the demo.

2. Create a namespace in Kubernetes to run HashiCorp Vault in. You can create a namespace using the following command:

```
$ kubectl create namespace vault
```

3. Create a file called `values.yaml` that contains the configuration settings for your Vault deployment:

```yaml
global:
  enabled: true
  imageRegistry: docker.io
  imagePullSecrets:
    - name: my-registry-secret

server:
  enabled: true
  image:
    repository: vault
    tag: 1.8.4
    pullPolicy: IfNotPresent
  env:
    VAULT_DEV_ROOT_TOKEN_ID: "myroot"
    VAULT_DEV_LISTEN_ADDRESS: "0.0.0.0:8200"
    VAULT_ADDR: "http://127.0.0.1:8200"
  extraVolumes:
    - type: secret
```

```
        name: vault-tls-cert
    extraVolumeMounts:
      - name: vault-tls-cert
        mountPath: /vault/tls
        readOnly: true
    ingress:
      enabled: true
      apiVersion: networking.k8s.io/v1
      pathType: Prefix
      hosts:
        - vault.local
      paths:
        - path: "/"
          pathType: Prefix
          backend:
            serviceName: vault
            servicePort: 8200
      tls:
        - secretName: vault-tls-cert
          hosts:
            - vault.example.com

  ui:
    enabled: true
    serviceType: ClusterIP

  replicaCount: 1

  persistence:
    enabled: true
    size: 10Gi
    storageClass: standard
```

This configuration enables high availability, allows ingress traffic, and sets an environment variable to point to the Vault server.

4. Install the HashiCorp Vault Helm chart. To do this, first, add the official HashiCorp Helm chart repository:

```
$ helm repo add hashicorp https://helm.releases.hashicorp.com
```

Then, install the chart using the following command:

```
$ helm install vault hashicorp/vault --namespace vault -f
values.yaml
```

This will install the HashiCorp Vault chart in the `vault` namespace using the configuration settings in the `values.yaml` file.

> **Important note**
>
> You need to have an Ingress Controller deployed and ensure that the vault namespace has access to that Ingress Controller. For internal secrets management, you won't need to deploy an ingress controller. However, to transmit secrets to a cloud service, or as a data feed to another remote service, the configured Ingress controller will come in handy.

Verify that the Vault server is running by checking its status using the following command:

```
$ kubectl exec -n vault -it vault-0 -- vault status
```

This should return the status of the Vault server, indicating that it is ready to use.

Once you have installed HashiCorp Vault, you can start creating and managing secrets using the Vault API and CLI. You can also integrate Vault with Kubernetes using the Vault Kubernetes authentication method to manage Kubernetes secrets.

Once Vault is installed in your Kubernetes environment, you can start with storing secrets. However, before that, we have to initialize and unseal the vault. You may follow these steps to unseal the vault:

1. First, you need to get the initial root token and unseal key. Run the following command to retrieve the unsealed key and route token:

    ```
    $ kubectl exec -it vault-0 -- /bin/sh
    $ vault operator init
    ```

2. Unseal the vault by running the following command three times for three different unseal keys:

    ```
    $ vault operator unseal
    ```

    When prompted, enter the first unseal key that you received in *step 1*.

3. Once you've entered the three unseal keys, Vault will be unsealed and ready to use. The output of the `unseal` command looks as follows:

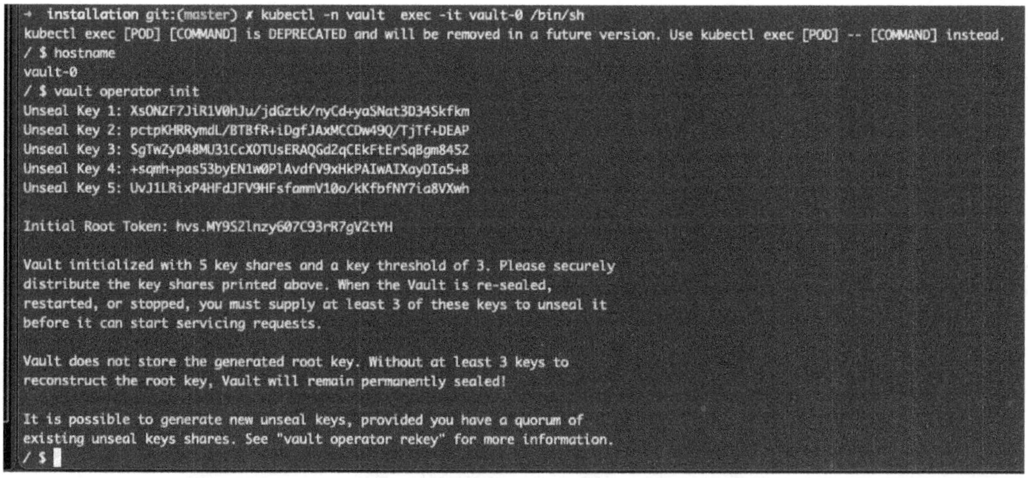

```
→ installation git:(master) x kubectl -n vault  exec -it vault-0 /bin/sh
kubectl exec [POD] [COMMAND] is DEPRECATED and will be removed in a future version. Use kubectl exec [POD] -- [COMMAND] instead.
/ $ hostname
vault-0
/ $ vault operator init
Unseal Key 1: XsONZF7JiR1V0hJu/jdGztk/nyCd+yaSNat3D34Skfkm
Unseal Key 2: pctpKHRRymdL/BT8fR+iDgfJAxMCCDw49Q/TjTf+DEAP
Unseal Key 3: SgTwZyD48MU31CcXOTUsERAQGdZqCEkFtErSqBgm845Z
Unseal Key 4: +sqmh+pas53byEN1w0PlAvdfV9xHkPAIwAIXayDIa5+B
Unseal Key 5: UvJ1LRixP4HFdJFV9HFsfammV10o/kKfbfNY7ia8VXwh

Initial Root Token: hvs.MY9SZlnzy607C93rR7gV2tYH

Vault initialized with 5 key shares and a key threshold of 3. Please securely
distribute the key shares printed above. When the Vault is re-sealed,
restarted, or stopped, you must supply at least 3 of these keys to unseal it
before it can start servicing requests.

Vault does not store the generated root key. Without at least 3 keys to
reconstruct the root key, Vault will remain permanently sealed!

It is possible to generate new unseal keys, provided you have a quorum of
existing unseal keys shares. See "vault operator rekey" for more information.
/ $ ▊
```

Figure 3.6 – Vault secret unseal

4. Now, you need to export the vault token so that you can access the Vault API. Run the following command to export the root token:

```
$ export VAULT_TOKEN=<root-token>
```

5. Verify that you can access the Vault API by running the following command:

```
$ vault status
```

If the response is sealed:false, you have successfully initialized and unsealed Vault:

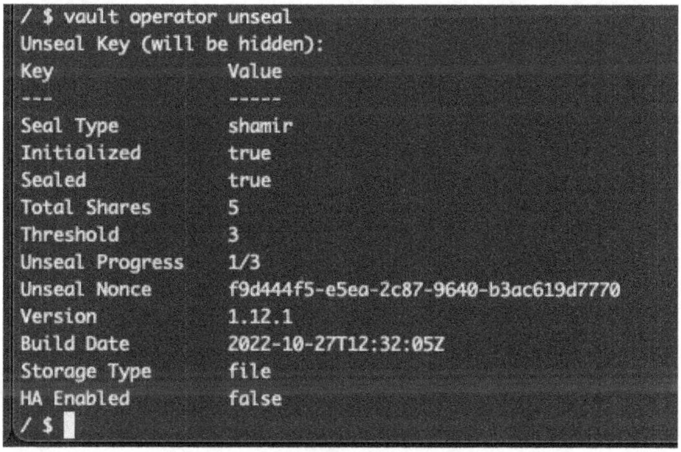

```
/ $ vault operator unseal
Unseal Key (will be hidden):
Key                   Value
---                   -----
Seal Type             shamir
Initialized           true
Sealed                true
Total Shares          5
Threshold             3
Unseal Progress       1/3
Unseal Nonce          f9d444f5-e5ea-2c87-9640-b3ac619d7770
Version               1.12.1
Build Date            2022-10-27T12:32:05Z
Storage Type          file
HA Enabled            false
/ $ ▊
```

Figure 3.7 – vault operator unseal

Congratulations! You have now initialized and unsealed HashiCorp Vault in your Kubernetes cluster. You can now start using Vault to manage your secrets and other sensitive information.

You can now create authentication and authorization; Vault supports a wide range of authentication and authorization methods, including Kubernetes RBAC, GitHub, and many others. Once you have identified an appropriate authentication method, you can configure it in Vault to control who can access it. You can now create secrets by using the Vault CLI or API and retrieve those secrets dynamically during runtime. The following figure explains how Vault functions with Kubernetes:

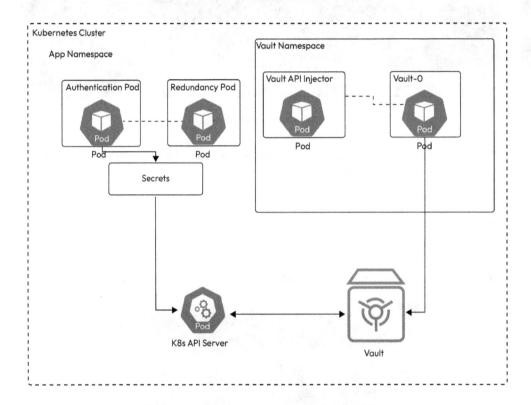

Figure 3.8 – Kubernetes Vault deployment architecture

How to create secrets in Vault

The process for creating secrets in Vault is as follows:

1. Authenticate to Vault by running the following command and entering your credentials:

    ```
    $ vault login <token>
    ```

> **Important note**
>
> You need to be *exec'd* into the Vault Pod before running the preceding command. You can exec into the Pod using the `kubectl exec` command.

Replace `<token>` with the token you obtained during the initialization and unsealing process.

2. Navigate to the path where you want to store the secrets. For example, if you want to store your secrets in a `secrets` path, run the following command:

```
$ vault kv put secrets/myapp/config username="myuser"
password="mypass"
```

This command creates a new secret in the `secrets/myapp/config` path and sets the `username` and `password` fields to `myuser` and `mypass`, respectively.

3. You can also create a JSON file containing the secret data and upload it to Vault using the `vault kv put` command. For example, say you have a file named `secrets.json` with the following content:

```
{
    "username": "myuser",
    "password": "mypass"
}
```

You can run the following command to upload the file to the `secrets/myapp/config` path:

```
$ vault kv put secrets/myapp/config @secrets.json
```

4. To retrieve the secrets stored in the `secrets/myapp/config` path, run the following command:

```
$ vault kv get secrets/myapp/config
```

This command retrieves the secrets stored in the `secrets/myapp/config` path and displays them on the screen.

> **Important note**
>
> Keep in mind that you need appropriate permissions to perform these operations, which can be set using Vault policies.

Summary

In this chapter, the focus was on the security considerations involved in cloud-native application development. The chapter provided an overview of cloud-native development and highlighted the differences between traditional and cloud-native app development. The DevOps model and how it fits into cloud-native architecture was also discussed.

The chapter explored the different security threats and attacks that can arise in cloud-native application development and introduced the best practices for integrating security into the development process. The OWASP Top 10 for Cloud-Native and the importance of not just focusing on shift-left and incorporating Dev-First Security were discussed.

This chapter also highlighted the security and development trade-off and discussed the benefits of incorporating supplemental security components into the development process. OWASP ASVS was also introduced as a tool for assessing the security posture of cloud-native applications.

Finally, the chapter provided an in-depth overview of secrets management and how to implement it strictly in the cloud-native space. HashiCorp Vault is used as an example to demonstrate how secrets management can be implemented in Kubernetes.

In the next chapter, we will incorporate everything we have learned so far to build a complete holistic Application Security program, which will include concepts such as secure coding and bug bounty.

Quiz

Answer the following questions to test your knowledge of this chapter:

- Kubernetes secrets provide a native solution to secret management, so why would organizations still seek to use HashiCorp Vault or any such third-party tools?

- How would you automate or scale the process of running image security scans as we learned how to do in the previous chapter and follow up with application security?

- Besides secrets, what other type of data would you, as a security engineer, be interested in preserving, and what steps could you take to do so?

- What are some of the security trade-offs to be mindful of when it comes to building an authentication and authorization service?

- How would you create a redundancy solution for secrets management using your Vault Pod?

Further reading

To learn more about the topics that were covered in this chapter, take a look at the following resources:

- `https://developer.hashicorp.com/vault/docs/secrets/kv`
- `https://owasp.org/www-pdf-archive/OWASP_Application_Security_Verification_Standard_4.0-en.pdf`
- `https://owasp.org/www-project-top-ten/`
- `https://www.ftc.gov/enforcement/refunds/equifax-data-breach-settlement`
- `https://www.forbes.com/sites/thomasbrewster/2019/07/24/deliveroo-accounts-hacked-and-sold-on-dark-web/?sh=3fe8114d42ff`
- `https://www.darkreading.com/attacks-breaches/capital-one-attacker-exploited-misconfigured-aws-databases`
- `https://kubernetes.io/docs/tasks/configmap-secret/managing-secret-using-kubectl/`

Part 2: Implementing Security in Cloud Native Environments

This part focuses on building an AppSec culture, threat modeling for cloud-native environments, securing the infrastructure, and cloud security operations. It also introduces DevSecOps practices, emphasizing IaC, policy as code, and CI/CD platforms. By the end of *Part 2*, you will have a comprehensive understanding of how to implement and manage security in a cloud-native environment.

This part has the following chapters:

4
Building an AppSec Culture

Building secure applications requires more than just fixing vulnerabilities found in the code. It requires a systematic approach that incorporates security into every stage of the development life cycle. An effective AppSec program is essential for any organization that wants to protect its data, customers, and reputation. In this chapter, we will cover the key components of building an AppSec program that is both effective and efficient. We will begin by discussing the importance of understanding your organization's security needs and goals. Next, we will explore the key elements of an effective AppSec program, including risk assessment, security testing, and security training. We will also cover the role of automation and the importance of building a culture of security within your organization. By the end of this chapter, you will have a clear understanding of how to build an AppSec program that meets the unique needs of your organization and provides a foundation for secure software development.

A successful AppSec program involves the collaboration of various stakeholders, including developers, security teams, and business leaders. It requires a shift in mindset, from reactive security measures to proactive security practices. It is not enough to simply rely on network security or traditional security measures. Organizations need to implement a comprehensive AppSec program that is tailored to their specific needs and aligned with their overall business goals.

We will discuss the importance of defining security policies and procedures that are specific to an organization's risk profile. We will also explore different techniques for risk assessment, which help identify potential vulnerabilities in the **software development life cycle** (SDLC). Additionally, we will cover the different types of security testing, including static analysis, dynamic analysis, and penetration testing. Finally, we will discuss the importance of security training and awareness programs and how to implement them effectively.

In this chapter, we're going to cover the following main topics.

- Overview of an AppSec program
- Building an effective AppSec program for cloud-native
- Developing policies and procedures
- Continuous monitoring and improvement

Technical requirements

The technical requirements for this chapter are as follows:

- Kubernetes v1.27
- A software development pipeline

Overview of building an AppSec program

An overview of building an AppSec program for cloud-native applications requires an understanding of the unique challenges presented by this environment. Cloud-native applications are built using microservices architecture and containers, which means they are highly distributed and dynamic. This presents a new set of security concerns that must be addressed to ensure that these applications are secure.

To build an effective AppSec program for cloud-native applications, it is important to start with a thorough risk assessment. This should include an analysis of the various components of the application, including the underlying infrastructure, the application code, and the network architecture. It is also important to consider the various security threats that may be present in a cloud-native environment, such as container vulnerabilities, API vulnerabilities, and data breaches.

Once the risks have been identified, the next step is to develop a comprehensive security plan that includes a range of security controls, including encryption, access controls, and intrusion detection. It is also important to implement a range of security testing methodologies, including static and dynamic code analysis, vulnerability scanning, and penetration testing. In addition, it is important to provide ongoing security training to developers and other stakeholders to ensure that everyone involved in the development process understands the importance of security and knows how to implement security best practices.

Finally, it is important to implement a range of automation tools and processes to help streamline the security process and ensure that security is incorporated into every stage of the development life cycle. This may include automated security testing tools, **Continuous Integration/Continuous Deployment** (CI/CD) pipelines, and other DevOps practices that help integrate security into the development process. By following these key steps, it is possible to build an effective AppSec program for cloud-native applications that provides a foundation for secure software development.

An AppSec program also helps in meeting compliance requirements and avoiding penalties, fines, and reputational damage associated with data breaches. It assures customers and stakeholders that the organization takes security seriously and has measures in place to safeguard their data. Moreover, an effective AppSec program can help reduce the cost of fixing security issues as early detection and remediation are typically less expensive than addressing security issues after an application has been deployed.

Understanding your security needs

Defining security requirements based on cloud-native architecture is the process of identifying, defining, and prioritizing security requirements that are specific to cloud-native environments. It involves understanding the unique security risks and challenges of cloud-native architecture and developing a set of security requirements that is tailored to address those risks and challenges.

These technologies enable greater flexibility, scalability, and agility in application development and deployment. However, they also introduce new security challenges, such as an increased attack surface area, container vulnerabilities, and distributed security controls.

From a security perspective, on reviewing the architecture, you should focus on not only the application code but also the infrastructure, network, data, and user access. This involves identifying and prioritizing security risks, defining security controls and policies, and implementing security technologies and processes that are appropriate for cloud-native environments.

An organization can define the requirements suited to it to build an AppSec program by following a systematic approach that takes into account its unique security needs, industry regulations, and compliance requirements. Here are some steps an organization can take to define its AppSec requirements:

1. **Understand the organization's security goals and objectives**: Before building an AppSec program, it's important to understand the security goals and objectives of the organization. This involves identifying the critical assets, systems, and data that need to be protected and determining the level of risk the organization is willing to accept.

2. **Conduct a risk assessment**: Conducting a risk assessment helps the organization identify potential vulnerabilities and threats to its systems and data. This process involves identifying potential threats, assessing the likelihood and impact of those threats, and determining the level of risk associated with each threat.

3. **Review industry regulations and compliance requirements**: Many industries have specific regulations and compliance requirements that organizations need to follow to ensure the security of their systems and data. It's important to review these regulations and requirements and ensure that the AppSec program is designed to meet these standards.

4. **Develop security policies and procedures**: Based on the organization's security goals, risk assessment, and compliance requirements, the organization should develop a set of security policies and procedures. These policies and procedures should outline the security measures that need to be implemented to protect the organization's systems and data.

5. **Determine the necessary security tools and technologies**: The organization should determine the necessary security tools and technologies that need to be implemented to support the AppSec program. This may include firewalls, intrusion detection systems, vulnerability scanners, and other security technologies.

6. **Establish a security culture**: Building a culture of security within the organization is essential to the success of the AppSec program. This involves training employees on security best practices, encouraging security awareness, and promoting a culture of continuous improvement.

This can further be broken down into the following pragmatic steps:

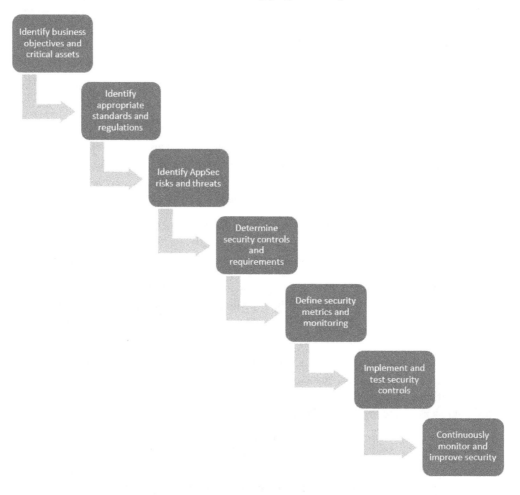

Figure 4.1 – AppSec program waterfall architecture

Let's talk about these steps in detail:

1. **Identify business objectives and critical assets**: The first step in building an AppSec program is to identify the business objectives and critical assets of your organization. This will help you understand which assets are most valuable and need to be protected, and which business objectives require a high level of security. This step is important because it helps you prioritize your security efforts and allocate resources where they are most needed.

2. **Identify appropriate standards and regulations**: Once you have identified your critical assets and business objectives, the next step is to identify the appropriate security standards and regulations that apply to your organization. These could include industry-specific regulations or government-mandated compliance requirements. Understanding the applicable standards and regulations will help you develop a comprehensive security program that meets all the requirements.

3. **Identify AppSec risks and threats**: With a clear understanding of your critical assets, business objectives, and applicable regulations, the next step is to identify the specific risks and threats that could impact your organization's security. This could include risks related to unauthorized access, data breaches, malware attacks, and more. By identifying these risks, you can develop a plan to mitigate them and protect your organization's assets.

4. **Determine security controls and requirements**: Once you have identified your risks and threats, the next step is to determine the security controls and requirements needed to protect your organization. This includes both technical and non-technical controls, such as access controls, firewalls, intrusion detection systems, and security awareness training for employees. The goal is to develop a comprehensive set of security requirements that will help mitigate the risks and threats identified in the previous step.

5. **Define security metrics and monitoring**: Once you have determined your security controls and requirements, the next step is to define the security metrics and monitoring that will be used to measure the effectiveness of your security program. This includes metrics related to incident response time, vulnerability management, and overall security posture. By establishing these metrics and monitoring processes, you can ensure that your security program is functioning as intended and adjust as necessary.

6. **Implement and test security controls**: With your security controls and requirements defined, the next step is to implement and test them to ensure that they are working effectively. This includes both technical testing, such as penetration testing and vulnerability scanning, as well as non-technical testing, such as social engineering testing. The goal is to ensure that your security controls are effective in mitigating the risks and threats identified earlier.

7. **Continuously monitor and improve security**: The final step in building an AppSec program is to continuously monitor and improve your security program. This includes ongoing testing and monitoring to ensure that your security controls are functioning as intended and are up to date with the latest threats and vulnerabilities. Additionally, you should regularly review and update your security controls and requirements to ensure that they remain effective in protecting your organization's critical assets and business objectives.

Identifying threats and risks in cloud-native environments

Cloud-native environments have unique characteristics and challenges that require a different approach to security compared to traditional data center environments. The following are some of the key steps you can take to identify threats and risks specific to cloud-native environments:

1. **Conduct threat modeling exercises**: Threat modeling is an effective way to identify potential threats and risks specific to cloud-native environments. This exercise involves creating a comprehensive diagram of the cloud-native environment, including all the components and data flows. With this diagram, you can identify potential threats and risks by considering each component's potential attack surface, entry points, and vulnerabilities.

2. **Review industry threat intelligence**: Keeping up to date with industry threat intelligence can help you identify emerging threats and risks specific to cloud-native environments. Many vendors and industry organizations offer regular threat intelligence reports that can help you stay informed and identify potential threats and risks.

3. **Use the Kubernetes threat matrix**: Kubernetes is a popular container orchestration system used in many cloud-native environments. The Kubernetes threat matrix is a community-driven project that provides an overview of the potential threats and risks specific to Kubernetes environments. This matrix includes detailed information on attack vectors, potential exploits, and recommended security controls to mitigate the risks.

4. **Consider compliance and regulatory requirements**: Cloud-native environments may have unique compliance and regulatory requirements that you must consider when identifying threats and risks. For example, the **General Data Protection Regulation** (**GDPR**) requires organizations to protect personal data, and failure to do so can result in significant financial penalties.

5. **Use automated security tools**: Automated security tools can help you identify potential threats and risks specific to cloud-native environments quickly. These tools can scan your cloud-native environment for vulnerabilities and potential misconfigurations that could be exploited by attackers.

6. To identify critical risks and threats, teams can follow a simple threat model:

Figure 4.2 – Simple threat model

In summary, you can use a combination of threat modeling, industry threat intelligence, the Kubernetes threat matrix, compliance and regulatory requirements, and automated security tools to identify potential threats and risks specific to your environment. Once identified, you can use this information to develop effective security controls and processes to mitigate the risks and protect your critical assets.

Bug bounty

One of the other strategic solutions would be to create a bug bounty program for the products that you want to secure. Bug bounty programs can be a valuable source for identifying product risks and threats in cloud-native environments. These are crowdsourced initiatives that invite security researchers and ethical hackers to identify security vulnerabilities in a company's product or application. These programs are often used by organizations to supplement their internal security testing efforts and to receive valuable feedback from external security experts.

The process typically begins with the organization setting up a program that outlines the scope of the program, the types of vulnerabilities that are eligible for rewards, and the reward amounts for different levels of vulnerabilities. The program is then made available to the public, and researchers are invited to participate by attempting to identify vulnerabilities in the organization's product or application. When a researcher identifies a vulnerability, they report it to the organization, which then assesses the vulnerability and determines whether it is valid and eligible for a reward. If the vulnerability is valid, the researcher is rewarded according to the program's rules.

One of the main advantages of bug bounty programs is that they allow organizations to leverage the expertise of a large pool of security researchers to identify vulnerabilities in their products or applications. This can help organizations identify security issues that they may not have been aware of, or that may have been missed during their internal security testing efforts. Additionally, these programs can be a cost-effective way for organizations to identify vulnerabilities as they only pay rewards for valid vulnerabilities that are identified.

However, there are also some disadvantages to bug bounty programs. For example, some organizations may be hesitant to implement a bug bounty program due to concerns about the potential for false positives, or the possibility of confidential information being exposed. Additionally, some security researchers may attempt to abuse the bug bounty program by submitting false vulnerabilities or attempting to exploit vulnerabilities without following the program's rules.

Despite these potential disadvantages, bug bounty programs can be a valuable tool for identifying product risks and threats in cloud-native environments. By supplementing internal security testing efforts with external expertise, organizations can improve the overall security of their products and applications and help ensure that they are more resilient to attacks and threats.

While bug bounty programs can be a powerful tool for organizations to use to identify and address potential vulnerabilities in their products or services, organizations must have an effective process in place for handling reported bugs to ensure that they are properly triaged and addressed.

When a bug is reported through a bug bounty program, the first step is typically triage by the engineering team. This involves assessing the severity and impact of the reported bug and determining the appropriate course of action. The triage process should be clearly defined and documented to ensure consistency and efficiency.

Once a bug has been triaged, the next step is to assign it to the appropriate team or individual for resolution. This may involve working with third-party vendors or external security researchers, depending on the nature of the bug and the resources available within the organization.

To ensure that bugs are fixed and closed promptly, it is important to establish clear and effective communication channels between the engineering team and the bug bounty program participants. This may involve setting expectations for response times and providing regular updates on the status of reported bugs.

Another time-effective practice is to prioritize the most critical and high-impact bugs first. This can help mitigate potential risks and minimize the impact of any vulnerabilities that may be present.

It is also important to establish clear guidelines for bug bounty payouts and ensure that they are consistent with industry standards. This can help incentivize participation in the bug bounty program and ensure that bugs are reported and addressed promptly.

Overall, a well-designed bug bounty program can be an effective way for organizations to identify and address potential product risks and threats. By establishing clear processes and communication channels, and prioritizing critical bugs, organizations can minimize risk and ensure the security and integrity of their products and services.

Evaluating compliance requirements and regulations

As organizations move toward cloud-native environments and adopt new technologies, they also face new compliance requirements and regulations that they must adhere to. Failure to comply with these regulations can result in significant financial and reputational damage, as well as legal penalties. Therefore, organizations need to evaluate these requirements and regulations and incorporate them into their AppSec program. Let's delve into more details regarding the elements that are factored in when analyzing compliance requirements for an organization.

Security service-level agreements (SLAs)

One of the first steps in evaluating compliance requirements is to determine the appropriate security **SLAs** for your organization. This involves understanding the level of security your organization requires and ensuring that the appropriate measures are in place to meet those requirements.

SLAs should be defined in terms of specific metrics, such as response times for security incidents or downtime due to security incidents. These metrics should be measurable, and their progress should be tracked to ensure that the organization is meeting its SLAs.

An SLA is a contractual agreement between a service provider and a customer that outlines the level of service the provider is expected to deliver. In the context of a cloud-native AppSec program, SLAs define the minimum level of service that customers can expect in terms of security. This includes things such as response times for security incidents, the availability of security resources, and measures for ensuring data privacy and confidentiality.

SLAs are essential in cloud-native AppSec programs for several reasons. Firstly, they provide a clear understanding of the security services and guarantees that are expected from the service provider. This is especially important in cloud environments where multiple parties are involved in providing and managing the infrastructure, applications, and data. With an SLA in place, all parties have a common understanding of their respective responsibilities and commitments.

Secondly, SLAs provide a mechanism for tracking and reporting on the performance of the security program. By defining specific security metrics and thresholds, organizations can monitor the effectiveness of their security controls and identify areas for improvement. This helps ensure that security is continuously optimized to meet evolving business needs and emerging threats.

Thirdly, SLAs play a crucial role in achieving security certifications such as FedRAMP. FedRAMP is a US government program that provides a standardized approach to assessing, authorizing, and continuously monitoring cloud services. To achieve FedRAMP compliance, **cloud service providers (CSPs)** must demonstrate that they meet stringent security requirements, including those related to SLAs. By implementing robust SLAs that meet FedRAMP requirements, organizations can demonstrate their commitment to delivering secure cloud services that meet the needs of government agencies.

When it comes to defining SLAs for a cloud-native AppSec program, there are several key considerations to keep in mind. These include the following:

- **Security metrics**: The SLA should define the specific security metrics that will be tracked and measured, such as incident response times, vulnerability remediation times, and data privacy controls.

- **Performance targets**: The SLA should define the minimum performance targets that the service provider is expected to meet for each security metric. This should be based on industry best practices, regulatory requirements, and the organization's specific risk profile.

- **Reporting**: The SLA should specify how security performance will be reported and communicated to customers. This should include regular reports, dashboards, and other mechanisms for providing transparency in security operations.

- **Remediation**: The SLA should define the process for addressing security incidents and vulnerabilities. This should include clear roles and responsibilities, escalation procedures, and timelines for remediation.

By defining robust SLAs that meet the needs of customers and regulatory requirements, organizations can ensure that their cloud services are secure, reliable, and meet evolving business needs.

Defining a security model

Once SLAs have been defined, the next step is to define a security model that will guide the engineering team in building and maintaining secure applications. This model should be based on industry standards and best practices, such as the OWASP Top 10, and should be tailored to the specific needs and requirements of the organization.

It also outlines the security controls and policies that should be in place to protect the organization's critical assets and business objectives. In a cloud-native environment, the security model should consider the unique risks and threats associated with this type of architecture.

One key consideration when drafting a security model for cloud-native environments is the use of microservices. Microservices are small, independently deployable services that work together to make up an application. Each microservice has its own set of security requirements and policies, and they all need to be coordinated and integrated to ensure the overall security of the application.

Another important consideration is the use of container orchestration platforms such as Kubernetes. These platforms provide a centralized way to manage containers and the applications running within them. The security model should account for the unique security challenges that come with using these platforms, such as securing the API server, managing access control, and implementing network segmentation.

When defining the security model for a cloud-native environment, it is also important to consider the various regulations and compliance requirements that may apply. For example, if the organization is operating in the healthcare industry, it may need to comply with the **Health Insurance Portability and Accountability Act (HIPAA)**. If it is operating in the financial industry, it may need to comply with the **Payment Card Industry Data Security Standard (PCI DSS)**. The security model should consider the specific requirements of these regulations and ensure that the necessary security controls are in place.

One way to ensure that the security model is effective is to conduct regular security audits and assessments. These assessments can help identify gaps and weaknesses in the security model and allow for adjustments and improvements to be made. Additionally, security certifications such as FedRAMP require organizations to demonstrate that they have a comprehensive security model in place that meets specific requirements. Having a well-defined security model can help streamline the process of achieving these certifications.

Consider an e-commerce company deploying its product in a cloud-native architecture while drafting the security model for such a product. The following key considerations should be factored in:

- User authentication and authorization:

  - Implement multi-factor authentication for all user accounts

  - Use OAuth 2.0/OpenID Connect to authenticate users and grant access to resources

  - Enforce strong password policies and regular password expiration

  - Implement **role-based access control** (**RBAC**) to ensure users have the appropriate level of access to resources

- Network security:

  - Implement secure communication protocols such as HTTPS and SSL/TLS for all data in transit

  - Use network segmentation to separate different tiers of the application and prevent lateral movement in case of a breach

  - Implement intrusion detection and prevention systems to monitor network traffic and detect potential attacks

- Infrastructure security:

  - Use cloud-native security tools such as Kubernetes RBAC and network policies to secure the underlying infrastructure

  - Implement regular vulnerability scans and penetration testing to identify and address potential vulnerabilities in the infrastructure

  - Encrypt all data at rest using encryption keys and secure key management practices, including key rotation and distribution

- Application security:

  - Implement a **secure SDLC** to ensure that security is built into the application from the beginning

  - Use automated testing tools such as static analysis and dynamic analysis to identify potential security vulnerabilities

  - Implement **runtime application self-protection (RASP)** to monitor the application in real time and detect potential attacks

- Compliance and auditing:

  - Implement logging and monitoring tools to track user activity and detect potential security incidents

  - Implement regular security audits and assessments to ensure compliance with industry standards and regulations such as the PCI DSS, HIPAA, and GDPR

  - Provide regular security training and awareness programs for all employees to ensure that they are aware of security policies and best practices

---

Important note

Drafting a security model isn't the same as conducting threat modeling. Even though they might seem very similar in the offset, the security model should be considered a best practice guide for developers to follow, whereas threat modeling is a result of the processes that have been drafted within the security model. Also, it is important to keep in mind that there is no silver bullet or a one-security-model-works-for-all solution; these are very much specific to each business unit within the organization, and a product security engineer should take responsibility for drafting and exercising it.

Tracking SLA progress

Once the SLAs and security model have been defined, it is important to track progress toward meeting these goals. This can be accomplished using **key performance indicators (KPIs)** and other metrics, which should be regularly monitored and reviewed.

It is also important to establish a reporting structure that allows for visibility into the status of the organization's security posture. This includes regular reports to management, stakeholders, and customers to communicate progress and ensure that everyone is aware of any potential security risks.

Tracking SLAs ensures that the security team is meeting the defined security objectives and expectations in a timely and effective manner. Besides providing a framework for measuring and tracking the performance of security controls and processes, they also help ensure that issues are addressed before they become critical.

SLAs are usually tied to specific security controls and processes. For example, an SLA might require that all high-priority security vulnerabilities be addressed within a certain time frame, or that all security incidents be responded to within a specific time frame. The SLA should also clearly define what is meant by "addressed" or "responded to" in each case.

One effective way to track SLAs is using automation tools such as Jira plugins. Jira can be integrated with pipeline security tools to provide real-time tracking and monitoring of SLAs. For example, a security vulnerability identified by a static analysis tool can automatically create a Jira ticket, which can then be assigned to the appropriate team member for resolution. The ticket can be tracked and monitored until the vulnerability is remediated and the SLA is met.

Jira automation also allows for the creation of custom workflows, which can be used to automate the SLA tracking process. For example, a workflow can be created to automatically escalate an issue if it has not been addressed within a certain time frame. This helps ensure that critical security issues are addressed promptly.

In addition to Jira automation, regular reporting and communication are also critical for tracking SLAs. Security teams should regularly report on their progress in meeting SLAs and communicate any issues or delays that may impact SLA performance. This allows stakeholders to stay informed and helps ensure that issues are addressed before they become critical.

Continuous improvement

Evaluating compliance requirements and regulations is not a one-time task, but an ongoing process that requires continuous improvement. This includes regular audits of the organization's security posture to identify any gaps or areas for improvement.

It is also important to stay up to date with changes to compliance requirements and regulations and to incorporate these changes into the AppSec program as needed. This can be accomplished through regularly training and educating the engineering team, as well as regularly reviewing relevant regulatory guidance.

Now that we understand what factors to consider for securing applications, let's look into actually bootstrapping an AppSec program for cloud engineering teams.

Building an effective AppSec program for cloud-native

Overall, the AppSec program should include not just technical tooling support and security expertise but also a set of soft skills. At the end of the day, if the engineering teams are not convinced that a security change needs to take effect, you will have a hard time convincing the team otherwise. Hence, it is vital to have the right tools in place to help support your case of making a security-related change within the development pipeline and imbibe security as a culture within the teams. In this section, we will do a deep dive into the nitty-gritty of each of the elements but from a 30,000-feet view. All security teams, while bootstrapping an AppSec program, should focus on the following key concepts:

- **Threat modeling**: Understanding the potential threats to the application and infrastructure is an important step in building a strong AppSec program. Threat modeling involves identifying and evaluating the risks and vulnerabilities in the system.

- **Code review**: Code review is a crucial part of any AppSec program. This involves reviewing the code for any security vulnerabilities or flaws that could be exploited by attackers.

- **Penetration testing**: Penetration testing involves testing the application and infrastructure for potential vulnerabilities by simulating a real-world attack. This helps identify any weaknesses in the system that could be exploited by attackers.

- **Security training**: Providing security training to developers and other staff is an important part of an effective AppSec program. This helps ensure that all members of the team understand the importance of security and are equipped with the knowledge and skills necessary to identify and mitigate potential threats.

- **Security testing automation**: Automating security testing processes can help identify vulnerabilities and other security issues more quickly and efficiently. This can include the use of tools such as static analysis, dynamic analysis, and software composition analysis.

- **Incident response**: Having an effective incident response plan in place is critical in the event of a security breach or other security incident. This involves having a clear process for identifying and responding to security incidents, as well as regularly testing and evaluating the incident response plan.

- **Ongoing monitoring and review**: Regularly monitoring and reviewing the AppSec program is important to ensure that it remains effective and up to date. This can include regular security assessments, vulnerability scans, and risk assessments to identify any potential areas for improvement or additional security measures.

Now, let's learn how a scalable AppSec program is implemented.

Security tools for software in development

By integrating security tools into the pipeline for software in development, organizations can identify and address vulnerabilities early in the development process, reducing the risk of security incidents in production environments.

Several types of security tools can be deployed within the pipeline for software in development, including the following:

- **Static application security testing (SAST) tools**: These tools analyze source code for potential security vulnerabilities without executing the code. SAST tools can identify a wide range of security issues, including injection attacks, buffer overflows, and insecure data storage.

- **Dynamic application security testing (DAST) tools**: These tools test running applications for security vulnerabilities. DAST tools can identify security issues that may not be apparent in the code, such as configuration errors or weaknesses in authentication mechanisms.

- **Interactive application security testing (IAST) tools**: These tools combine elements of SAST and DAST, analyzing the code and testing the application in real time. IAST tools can provide more accurate results than either SAST or DAST tools alone.

- **Software composition analysis (SCA) tools**: These tools identify and track open source components used in software development. SCA tools can help organizations ensure that open source components are up to date and free of known vulnerabilities.

- **Container security tools**: These tools help organizations identify vulnerabilities in container images and monitor containers in runtime. Container security tools can detect container configuration issues, unauthorized access, and anomalous behaviors.

- **Infrastructure as Code (IaC) security tools**: These tools help organizations ensure the security of their IaC templates, which define the cloud infrastructure, network architecture, and resources that support their applications.

By deploying a combination of these tools within the pipeline for software in development, organizations can improve their overall security posture and reduce the risk of security incidents. It's important to note that no tool can guarantee 100% security and that security tools should be used in conjunction with other security measures, such as security training for developers and robust incident response plans.

> **Important note**
>
> When you're considering deploying these security tools at scale for multiple products that have a huge code base, it is crucial to think about the turnaround time and the repercussions of that based on the SLAs and the engineering team's signoff. If the insecure code would be a promotion blocker, then activating the tools should be considered pre-commit and post-commit.

To minimize the impact of promoting code, security tools can be integrated at various stages of the commit process. For example, for the pre-commit stage, developers can utilize tools such as pre-commit hooks to run security checks before the code is even committed. In the post-commit stage, tools such as Snyk, SonarQube, and Twistlock can be integrated to scan the code base for vulnerabilities and provide feedback to the developers.

Similarly, in the pre-merge stage, developers can use tools such as Checkov, CloudFormation Guard, and Terraform Sentinel to conduct IaC security scans. This ensures that the infrastructure code is secure and does not introduce any vulnerabilities. By integrating these tools into the pipeline, developers can identify and fix security issues early in the development process, reducing the risk of vulnerabilities being deployed to production. Finally, IaC security scans can be run in the pre-merge stage of a pull request to ensure that any code changes do not introduce security issues into the cloud infrastructure.

By running IaC security scans in these stages, the time taken to run the scans is minimized, and the overall security of the cloud infrastructure can be improved:

- **Pre-commit**: Pre-commit checks are performed on a developer's workstation before the code is committed to the repository. These checks typically include linting, syntax checking, and unit testing. Security tools that can be deployed at this stage include static code analysis tools, which can identify security vulnerabilities in the code as it is being written. These tools can provide feedback to developers about issues such as injection flaws, **cross-site scripting (XSS)** vulnerabilities, and other common coding errors. By identifying these issues early in the development process, developers can address them before the code is committed to the repository, saving time and effort in the long run.

- **Post-commit**: Post-commit checks are performed on the code that has been committed to the repository. These checks typically include dynamic code analysis, which involves testing the application by interacting with it as a user would. Security tools that can be deployed in this stage include **web application firewalls (WAFs)**, which can monitor and filter traffic to the application to identify and block potential attacks. Another example of a post-commit security tool is a vulnerability scanner, which can scan the application for known vulnerabilities and provide detailed reports on any issues that are found.

- **Pre-merge**: Pre-merge checks are performed on the code that is being merged into the main code base. These checks typically include integration testing, which involves testing the application in the context of the overall system. Security tools that can be deployed in this stage include IaC security scanners, which can identify security issues in the configuration files used to deploy the application. By identifying misconfigurations and other issues before the code is merged, developers can avoid introducing security issues into the main code base.

To ensure that these security tools do not slow down the development process, it is important to automate their deployment and integration with the pipeline. One way to achieve this is by using platforms such as Jenkins/Codefresh or other similar CI/CD tools, which can be configured to automatically trigger security scans at different stages of the pipeline. For example, a pre-commit scan can be triggered automatically when a developer submits a pull request, while a post-commit scan can be triggered automatically when the code is merged into the main branch. By automating these processes, developers can focus on writing code while the security tools run in the background, providing feedback on potential security issues in real time:

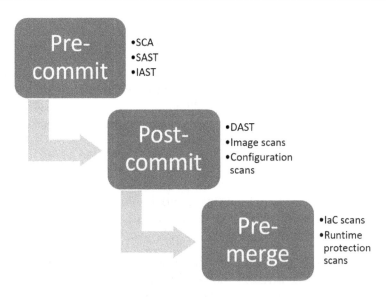

Figure 4.3 – Promotion cycle for a PR

In addition to automation, you would also have to track the results of these security scans over time to identify trends and measure progress. This can be done by setting security SLAs and tracking key metrics such as the number of vulnerabilities detected and the time to remediation. By measuring these metrics, organizations can identify areas for improvement and adjust their AppSec program accordingly.

> **Important note**
>
> It should also be considered that if the business unit is running QA scans, then the scans should be triggered in parallel to the security scans. While it might cost a marginal amount more to multi-thread by most vendors, it is always beneficial to not have security scans as a blocker for the promotion of code since this might cause a delay to take the application to market.

If the scan results show considerably high/severe security misconfigurations, the security engineer should consult the developer to fix the code as soon as the ticket is created and the SLA clock starts to tick. This is to ensure that the findings are not lost when multiple tickets are created.

Threat modeling

Threat modeling involves identifying potential threats to and vulnerabilities within the system and taking proactive measures to mitigate or eliminate them. One important aspect of threat modeling in cloud-native environments is to identify the different attack surfaces that exist within the system. Attack surfaces can include anything from exposed APIs to vulnerable container images. Understanding the different attack surfaces is critical to identifying potential threats and vulnerabilities and deciding which security controls are necessary to mitigate them.

Another important consideration when threat modeling in cloud-native environments is to account for the dynamic nature of the system. Microservices, for example, can be spun up and down quickly, and containers can be migrated between hosts. This means that the attack surface can change rapidly, making it essential to conduct frequent threat modeling exercises to ensure that the security controls are up to date and effective.

Additionally, it is important to consider the specific threat actors that may target the system. These could include insider threats, external hackers, or even automated bots. Understanding the motivations and tactics of these threat actors is critical to developing an effective threat model and selecting appropriate security controls.

One approach to threat modeling in cloud-native environments is to use a tool such as OWASP Threat Dragon. This open source tool allows teams to create visual diagrams of their system architecture and identify potential threats and vulnerabilities. The tool also provides guidance on selecting appropriate security controls based on the identified threats.

While we will discuss this in detail in *Chapter 5*, let's understand why it is an important step for an AppSec program. Integrating threat modeling into the AppSec culture of a company requires a combination of education, tools, and processes. AppSec teams can help educate developers on the importance of threat modeling and how it fits into the larger security landscape. They can also provide training on how to conduct threat modeling exercises and best practices for integrating it into the development process.

Tools and technologies can also play a key role in integrating threat modeling into the AppSec culture. There are several threat modeling tools available that can help teams identify potential risks and vulnerabilities. These tools can be integrated into the development process, providing continuous feedback to developers on potential security issues. Additionally, integrating threat modeling into the company's DevOps pipeline can help ensure that security is built into the entire SDLC.

Threat modeling can also be used to facilitate collaboration between development and security teams, promoting a culture of shared responsibility for security. By involving developers in the threat modeling process, they gain a better understanding of the security risks associated with their code and become more invested in creating secure applications. This collaboration helps break down silos between teams, leading to more effective communication and faster response times to security issues.

Several steps need to be followed to conduct threat modeling:

1. **Define the application scope**: The first step in threat modeling is to define the scope of the application. This involves identifying the critical assets, data flows, and potential attack surfaces that are relevant to the application. This can be done through documentation, interviews with stakeholders, and examining the architecture of the application. For example, in an e-commerce application, the scope would include customer data, payment processing, and third-party integrations.

2. **Create a data flow diagram**: Once the scope has been defined, the next step is to create a **data flow diagram (DFD)** that illustrates how data moves through the application. This can help identify potential attack surfaces and points of entry for attackers. For example, in an e-commerce application, the DFD would show how customer data flows from the user interface to the database and payment gateway.

3. **Identify threats**: With the scope and DFD in place, the next step is to identify potential threats to the application. This can be done through brainstorming and examining the various attack vectors that could be used to compromise the application. For example, in an e-commerce application, threats could include SQL injection, XSS, and payment fraud.

4. **Rank threats**: Once threats have been identified, they should be ranked based on their likelihood and impact. This can be done using a risk matrix, which assigns a risk score based on the likelihood and impact of each threat. For example, a high-likelihood and high-impact threat, such as payment fraud, would receive a higher risk score than a low-likelihood and low-impact threat, such as a user forgetting their password.

5. **Identify vulnerabilities**: After identifying and ranking threats, the next step is to identify potential vulnerabilities in the application that could be exploited by attackers. This can be done through code review, static analysis, and penetration testing. For example, in an e-commerce application, vulnerabilities could include unsecured API endpoints or weak authentication mechanisms.

6. **Identify countermeasures**: Once vulnerabilities have been identified, the next step is to identify countermeasures that can be used to mitigate the risk of each threat. This can include implementing secure coding practices, using encryption, and implementing access controls. For example, in an e-commerce application, countermeasures could include implementing two-factor authentication, encrypting customer data, and using a WAF.

7. **Review and repeat**: Finally, the threat modeling process should be reviewed and repeated regularly to ensure that the application remains secure over time. This can include reviewing the threat model after major updates to the application or changes have been made to the cloud-native environment:

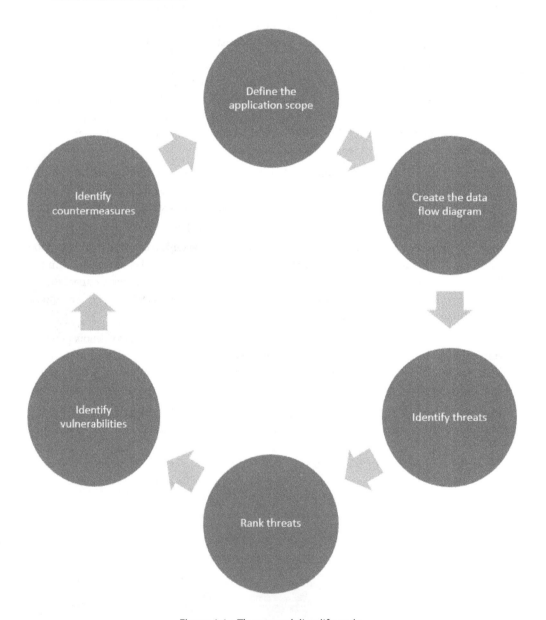

Figure 4.4 – Threat modeling life cycle

Threat modeling is an essential part of the AppSec culture within any organization. It helps identify potential vulnerabilities and threats in the application before they can be exploited by attackers. By conducting threat modeling regularly, organizations can ensure that their applications remain secure and are better prepared to respond to new threats.

Real-world examples of the importance of threat modeling include the Equifax data breach in 2017, which exposed the personal information of over 140 million people. This breach was attributed to a vulnerability in the Apache Struts framework that could have been identified and mitigated through threat modeling. Another example is the Capital One data breach in 2019, which exposed the personal information of over 100 million people. This breach was caused by a misconfigured firewall, which could have been identified through a thorough threat modeling process.

Providing security training and awareness to all stakeholders

One of the key aspects of building an AppSec culture across an organization is providing security awareness and training to all stakeholders involved in developing and deploying applications. In this section, we will discuss the importance of security awareness and training and provide a comprehensive guide on how the AppSec engineering team can effectively provide such training.

Stakeholders in an application security program can include developers, DevOps engineers, project managers, business owners, and end users. Each stakeholder has a different level of understanding and expertise in security, and it is the responsibility of the AppSec engineering team to ensure that everyone is on the same page when it comes to security.

The importance of security awareness and training cannot be overstated. In addition to improving the security posture of the organization, security training also helps create a culture of security within the company. When everyone is aware of the risks and the role they play in mitigating them, it becomes easier to identify and address security issues.

When designing a security awareness and training program, it is important to define the objectives and end goals. Some common objectives of a security training program are as follows:

- Educating stakeholders about the importance of security
- Teaching stakeholders how to identify and mitigate security risks
- Encouraging stakeholders to report security incidents or vulnerabilities

- Creating a culture of security within the organization:

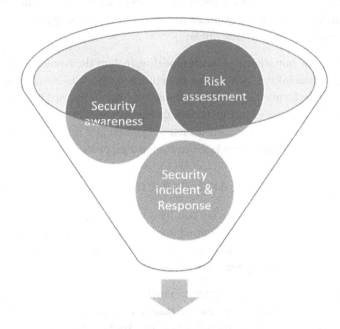

Security awareness program

Figure 4.5 – Components of security awareness

The AppSec engineering team can use various tools and techniques to deliver security training. Some common methods include online courses, webinars, workshops, and hands-on training. It is important to tailor the training to the specific needs of the stakeholders involved. For example, developers may require more technical training on secure coding practices, while project managers may require a higher-level overview of security risks and how they can be addressed.

In addition to traditional training methods, engineering teams can also use gamification to make the training more engaging and interactive. For example, they can create security challenges or capture-the-flag events, which encourage stakeholders to apply what they have learned in a real-world scenario.

To measure the effectiveness of the training program, the AppSec engineering team can use various metrics, such as the number of reported security incidents or vulnerabilities, the percentage of stakeholders who complete the training, or the reduction in the number of security incidents over time.

When designing the training program, it is also important to consider the different learning styles of the stakeholders. Some may prefer visual aids such as diagrams or videos, while others may prefer hands-on training. Considering a mix of techniques ensures that the training is effective for all stakeholders.

It is also important to keep the training up to date with the latest security risks and best practices. You should conduct regular reviews of the training program and make updates as necessary to ensure that stakeholders are equipped with the latest knowledge and skills.

In addition to the points mentioned previously, there are a few more aspects that could be considered when providing security awareness and training to stakeholders:

- **Role-specific training**: The training should be tailored to the specific roles and responsibilities of each stakeholder. For example, developers may need training on secure coding practices, while business analysts may need training on data protection regulations.

- **Continuous training**: Security training should not be a one-time event but rather a continuous process. New threats and vulnerabilities are constantly emerging, and stakeholders need to be kept up to date on the latest security best practices.

- **Simulation exercises**: Realistic simulations of security incidents can help stakeholders better understand the risks and consequences of security breaches. These simulations can also provide an opportunity for stakeholders to practice their incident response procedures.

- **Metrics and evaluation**: It's important to track the effectiveness of the security awareness and training program by measuring the improvement in stakeholder knowledge and behavior. Regular evaluations can help identify areas for improvement and ensure that the training program is meeting its goals.

- **Incentives**: Providing incentives for stakeholders who demonstrate good security practices can help promote a culture of security within the organization. For example, developers who identify and report security vulnerabilities could be recognized and rewarded for their efforts.

- **Collaboration with other teams**: The AppSec engineering team should collaborate with other teams within the organization to ensure that security is integrated into all aspects of the SDLC. For example, the AppSec team could work with the operations team to ensure that security monitoring tools are in place and that security incidents are responded to promptly.

Overall, providing effective security awareness and training requires a proactive approach that is tailored to the needs of each stakeholder group, is continuously updated to reflect the latest security best practices, and is regularly evaluated to ensure that it is achieving its goals.

Developing policies and procedures

Establishing a governance model for AppSec program management involves defining a structured framework for managing the organization's application security program. This framework is used to ensure that application security is managed and maintained consistently across the organization and that it aligns with the overall business objectives.

A governance model for AppSec program management includes establishing policies, procedures, and standards that govern the entire application security program. It defines the roles and responsibilities of various stakeholders involved in the program, including the security team, development team, project managers, and executive management. It also includes the definitions of KPIs and metrics that are used to measure the effectiveness of the program.

The governance model ensures that the application security program is managed in a consistent and structured manner and that there is clear accountability and responsibility for its implementation. It also provides a means to evaluate and continuously improve the program by identifying areas of weakness and implementing corrective measures.

An effective governance model for AppSec program management includes the following key elements:

- **Policies, procedures, and standards**: A set of policies, procedures, and standards that is defined to govern the entire application security program. These documents provide a clear set of guidelines that is followed by all stakeholders involved in the program.

- **Roles and responsibilities**: The governance model defines the roles and responsibilities of various stakeholders involved in the program. This includes the security team, development team, project managers, and executive management.

- **KPIs and metrics**: The governance model includes the definitions of KPIs and metrics that are used to measure the effectiveness of the program. These metrics should be aligned with the overall business objectives and should be used to continuously evaluate and improve the program.

- **Risk management**: The governance model should include a risk management framework that is used to identify, assess, and mitigate risks associated with the application security program. This includes a process for risk identification, risk assessment, risk mitigation, and risk monitoring.

- **Continuous improvement**: The governance model should provide a framework for continuously improving the application security program. This includes a process for evaluating the effectiveness of the program and implementing corrective measures to improve its effectiveness.

Incident response and disaster recovery

Developing an incident response and disaster recovery runbook involves a set of predefined procedures that helps identify, analyze, contain, and recover from security incidents or unexpected outages.

To develop an effective incident response and disaster recovery runbook, there are several key considerations that organizations should keep in mind. These include the following:

- **Defining the scope**: The first step in developing a runbook is defining the scope of the document. This includes identifying the systems, applications, and data that are critical to the organization and need to be protected in case of a security incident or disaster.

- **Outlining the incident response process**: The incident response process should be outlined in the runbook, including roles and responsibilities, communication protocols, and escalation procedures. This will ensure that everyone involved in the response knows their roles and responsibilities and can work together efficiently.

- **Identifying potential threats**: Organizations should identify potential threats to their systems, applications, and data and outline the steps that should be taken to prevent or mitigate these threats. This can include vulnerability assessments, penetration testing, and regular security audits.

- **Defining disaster recovery procedures**: The disaster recovery plan should include procedures for restoring systems, applications, and data in case of an outage or other disaster. This should include backup and recovery procedures, failover strategies, and testing protocols.

- **Testing and updating the runbook**: Once the runbook has been developed, it is important to test it regularly to ensure that it is effective and up to date. Testing can include tabletop exercises, simulations, and live tests. Any updates or changes to the runbook should be documented and communicated to all stakeholders.

When developing an incident response and disaster recovery runbook for a cloud-native environment, there are additional considerations to keep in mind. Let's take a look:

- **CSP responsibilities**: Organizations should understand the responsibilities of their CSPs in the event of an incident or outage. This includes understanding the CSP's disaster recovery procedures and SLAs.

- **Multi-cloud environments**: Organizations that operate in multi-cloud environments should consider the unique challenges that come with managing and securing data across multiple cloud platforms. This includes ensuring that the runbook applies to all cloud environments and that communication protocols are established across all cloud providers.

- **Automation**: Automation can play a critical role in incident response and disaster recovery in cloud-native environments. Organizations should consider using automated tools to monitor their systems and applications, detect potential threats, and respond to incidents in real time.

By following these key considerations and considering the unique challenges of cloud-native environments, organizations can ensure that they are well prepared to respond to security incidents and unexpected outages. Now, let's learn about the different components of a cloud security policy.

Cloud security policy

A cloud security policy is a set of guidelines and principles that an organization establishes to ensure the confidentiality, integrity, and availability of its data and applications in a cloud environment. The policy defines the rules and procedures for protecting an organization's information assets and outlines the controls and measures required to safeguard them against potential security threats.

A framework for defining a cloud security policy typically includes the following steps:

1. **Risk assessment**: Conduct a thorough assessment of the potential risks and threats to the organization's cloud environment. This involves identifying the types of data and applications that will be stored in the cloud and the potential vulnerabilities associated with each.

2. **Define objectives**: Establish the specific objectives and goals of the cloud security policy. This may include ensuring compliance with regulatory requirements, protecting against data breaches, and safeguarding intellectual property.

3. **Define controls**: Identify the specific controls and measures that will be put in place to mitigate risks and protect the organization's data and applications in the cloud. This may include network security measures, access controls, encryption, and data backup and recovery procedures.

4. **Develop the policy**: Draft the cloud security policy document. This should clearly outline the organization's security objectives, controls, and procedures.

5. **Implement the policy**: Put the cloud security policy into action by deploying the necessary security controls and measures across the organization's cloud environment.

Once a cloud security policy has been established, it is important to measure its effectiveness through regular audits and testing. Existing benchmarks for measuring the effectiveness of a cloud security policy include the Cloud Security Alliance's Cloud Controls Matrix, which provides a comprehensive list of security controls that organizations can use to assess their cloud security posture.

Organizations must also ensure that all stakeholders, including employees, contractors, and third-party service providers, are aware of the policy and trained on the specific security measures and procedures required to comply with it. It is also pretty common to consider using cloud security automation tools to enforce security policies and ensure that all cloud infrastructure and applications are configured and maintained as per the policy.

In following a comprehensive framework for developing and implementing a cloud security policy, organizations can mitigate the risks associated with cloud computing and ensure that their data and applications remain secure. Regular audits and testing can help measure the effectiveness of the policy, while training and automation tools can ensure that all stakeholders are aware of the policy and compliant with its requirements.

Organizations need to be mindful of various privacy laws and regulations when developing cloud security policies. Here are some of the privacy laws that organizations should consider:

- **The GDPR**: The GDPR is a regulation that came into effect in 2018 and applies to all organizations that process the personal data of EU citizens. This regulation requires organizations to implement appropriate technical and organizational measures to ensure the security of personal data.

- **The California Consumer Privacy Act (CCPA)**: The CCPA is a privacy law that applies to organizations that do business in California and collect the personal data of California residents. The law requires organizations to provide consumers with certain rights, such as the right to access, delete, and opt out of the sale of their data.

- **The HIPAA**: The HIPAA is a privacy law that applies to healthcare organizations and their business associates. The law requires organizations to implement appropriate safeguards to ensure the confidentiality, integrity, and availability of protected health information.

- **The PCI DSS**: The PCI DSS is a set of security standards that applies to organizations that accept payment cards. These standards require organizations to implement various technical and organizational measures to protect payment card data.

- **The Children's Online Privacy Protection Act (COPPA)**: The COPPA is a privacy law that applies to organizations that collect personal information from children under the age of 13. The law requires organizations to obtain parental consent before collecting personal information from children.

To comply with these privacy laws, organizations need to ensure that their cloud security policies include appropriate measures to protect personal data, such as implementing encryption, access controls, and data retention policies to protect personal data. Additionally, organizations need to ensure that their policies provide consumers with the necessary rights and disclosures required by these privacy laws.

To measure the effectiveness of these policies, organizations can use various benchmarks and frameworks, such as ISO 27001, NIST Cybersecurity Framework, and CIS Controls. These frameworks provide guidelines and best practices for developing and implementing effective security policies.

Identity and access management policies

Identity and access management (IAM) policies are critical components of any security program. These policies define the permissions and access controls for users, devices, and applications that access an organization's resources. IAM policies ensure that only authorized users can access sensitive data, applications, and systems. In this section, we will discuss what IAM policies are, why they are important, the framework to define the policy, existing benchmarks for measuring policy effectiveness, and what organizations can do to implement them.

IAM policies provide a framework for managing access to resources in the cloud. They help organizations establish who has access to what and what actions they can perform. IAM policies allow for specific rules that govern access to resources to be created. These policies can be granular, providing a high degree of control over what users can do and which resources they can access. IAM policies can be used to manage permissions for users, groups, and roles. They can also be used to manage access to cloud services, applications, and data.

A properly defined IAM policy can help organizations enforce the principle of least privilege, which means that users are only given access to the resources they need to perform their jobs. IAM policies can also help organizations comply with regulations such as the GDPR, HIPAA, and PCI DSS.

The framework for defining IAM policies involves several steps. First, organizations need to identify their resources and the users who need access to them. Next, they need to define the access control rules that govern who has access to what resources. This involves identifying the permissions required to access the resources and mapping those permissions to users, groups, and roles. Finally, organizations need to test and refine their policies to ensure that they are effective and do not create unnecessary access controls.

Measuring the effectiveness of IAM policies can be challenging. One benchmark that is often used is the principle of least privilege. This principle states that users should only have the permissions they need to perform their jobs. By following this principle, organizations can reduce the risk of unauthorized access to sensitive data and applications. Another benchmark is the number of security incidents that occur. A well-defined IAM policy can help reduce the number of security incidents by ensuring that only authorized users have access to resources.

It is efficient to implement IAM policies using a variety of tools and techniques. Many cloud providers offer IAM services that can be used to manage access to cloud resources. These services typically provide a range of features, including authentication, authorization, and access control. Organizations can also use third-party tools to manage IAM policies. These tools typically provide more granular control over access to resources and can be used to enforce policies across multiple cloud providers.

Continuous monitoring and improvement

Incorporating continuous monitoring involves leveraging automated tools and techniques to monitor and track the security posture of an organization's software applications and infrastructure continuously.

One of the key components of continuous monitoring is vulnerability scanning. This involves the use of tools such as static and dynamic analysis to identify potential security vulnerabilities in software code and infrastructure. These scans should be integrated into the DevOps pipeline, allowing for quick identification and remediation of any security issues.

In addition to vulnerability scanning, continuous monitoring also involves collecting and analyzing security-related data from various sources, such as logs, metrics, and alerts. This data can then be analyzed to identify potential security issues and track the effectiveness of existing security controls.

The first step is for organizations to establish a baseline for their security posture. This baseline should include a set of key metrics that will be tracked over time to measure the effectiveness of the AppSec program.

Once the baseline has been established, organizations should identify areas where improvements can be made and develop a plan to address these issues. This plan should include implementing new security controls, modifying existing controls, and integrating new monitoring tools and techniques.

To ensure that the continuous monitoring and improvement process is effective, organizations should regularly review and analyze the data that's collected from the monitoring tools. This analysis should be used to identify trends and patterns that can be used to improve the overall security posture of the organization.

Incorporating continuous monitoring and improvement into the DevSecOps process requires a cultural shift toward dev-first security. This means that security is incorporated into every step of the development process, from design to deployment. It also involves collaboration between developers and security professionals to ensure that security is not an afterthought but is integrated into the development process from the start.

Summary

An effective AppSec program for cloud-native environments should incorporate threat modeling, vulnerability management, secure development training, governance, incident response, disaster recovery, cloud security policies, and IAM policies.

Threat modeling helps identify potential security risks and develop countermeasures. Vulnerability management involves continuously scanning for and mitigating vulnerabilities in the code. Secure development training should be provided to all stakeholders in the organization, including developers, QA engineers, and product owners.

Governance involves establishing policies and procedures for managing the AppSec program, including incident response and disaster recovery plans. Cloud security policies should be established to define the security posture of the organization and ensure compliance with relevant laws and regulations. IAM policies help ensure that users and systems have the appropriate access to resources.

Continuous monitoring and improvement are important for maintaining the security of a cloud-native environment. Organizations can implement automated testing and scanning tools as part of a DevSecOps approach to catch vulnerabilities early in the development process. They can also establish metrics and benchmarks for measuring the effectiveness of the AppSec program and identify areas for improvement.

Incorporating security into the DevOps process involves making security a priority from the start of the development process, rather than as an afterthought. This requires a shift in mindset and culture to ensure that all stakeholders understand the importance of security and are committed to implementing it throughout the development life cycle.

Overall, a robust AppSec program for cloud-native environments involves a comprehensive approach that addresses all aspects of security, including threat modeling, vulnerability management, secure development training, governance, incident response, disaster recovery, cloud security policies, and IAM policies. By incorporating security into the DevOps process and implementing continuous monitoring and improvement, organizations can build and maintain a secure cloud-native environment.

Threat modeling is a crucial aspect of any comprehensive AppSec program. By identifying potential security threats and vulnerabilities early in the development process, organizations can save time and resources while ensuring the security of their applications. In this next chapter, we will delve deeper into the topic of threat modeling, exploring its key concepts, methodologies, and best practices. We will also examine how threat modeling can be used in the context of cloud-native environments and provide real-world examples to illustrate its importance. By the end of the next chapter, you will have a clear understanding of what threat modeling is, how it works, and why it is essential for building secure applications in today's digital landscape.

Quiz

Answer the following questions to test your knowledge of this chapter:

- What is the goal of an AppSec program?
- What are some common tools that are used in AppSec programs?
- How can security awareness and training be provided to stakeholders in an organization?
- What is the purpose of establishing a governance model for AppSec program management?
- What are some common policies and procedures that should be implemented in an AppSec program for cloud-native environments?

Further readings

To learn more about the topics that were covered in this chapter, take a look at the following resources:

- `https://csrc.nist.gov/publications/detail/sp/800-53/rev-5/final`
- `https://cloudsecurityalliance.org/research/guidance/`
- `https://landing.google.com/sre/books/`
- `https://owasp.org/www-project-threat-dragon/`
- `https://www.microsoft.com/en-us/securityengineering/sdl`
- `https://www.cisecurity.org/controls/`

5

Threat Modeling for Cloud Native

In this chapter, we will be exploring a critical aspect of cloud-native security—**threat modeling**. Threat modeling is a systematic approach to identifying, quantifying, and prioritizing potential threats and vulnerabilities in a cloud-native environment.

So far in this book, we have covered the foundations of cloud-native, application security, system security, and developing an application security culture. In this chapter, we will bring all these concepts together and apply them to the process of threat modeling. You will learn how to identify potential threats, assess their impact and likelihood, and develop mitigation strategies to protect your cloud-native environment.

By the end of this chapter, you will have a comprehensive understanding of how to perform threat modeling for cloud-native environments and how to use the information gathered to make informed decisions about security risks.

In this chapter, we're going to be covering the following topics:

- Developing an approach to threat modeling
- Developing a threat matrix
- Threat modeling frameworks
- A Kubernetes threat matrix

Technical requirements

We will be using the following tools in this chapter:

- A Kubernetes threat matrix
- Threat modeling frameworks—**STRIDE**, **PASTA**, and **LINDDUN**

Developing an approach to threat modeling

In this section, we will delve into developing an approach to threat modeling specifically designed for cloud-native environments.

As organizations transition to cloud-native architectures, they need to adapt their threat modeling processes to account for the unique challenges and opportunities these environments present. In this overview, we'll explore the importance of threat modeling, key concepts, and essential steps to create an effective threat model for cloud-native environments.

An overview of threat modeling for cloud native

This overview will cover the importance of threat modeling in cloud-native environments, discuss the unique challenges, and provide an outline for a successful cloud-native threat modeling process.

Importance of threat modeling in cloud-native environments

As more organizations embrace cloud-native technologies and methodologies, understanding the potential threats to these environments becomes increasingly crucial. Cloud-native applications often rely on distributed architectures, microservices, and containerization, leading to a complex web of interdependencies and potential attack surfaces. Additionally, the dynamic and ephemeral nature of cloud-native environments increases the difficulty in tracking and securing assets.

Threat modeling provides a structured approach to identifying and prioritizing risks, allowing organizations to make informed decisions about security investments and risk mitigation strategies. By proactively identifying potential threats, organizations can address vulnerabilities before they become exploited, reducing the likelihood of a security incident and minimizing the potential impact.

Unique challenges in cloud-native threat modeling

Threat modeling in cloud-native environments poses unique challenges that require a tailored approach. Some of these challenges include the following:

- **Complexity and interdependency**: Cloud-native environments often consist of numerous interconnected components, such as microservices, containers, and APIs. This complexity makes it challenging to map out the entire environment and understand potential attack surfaces and interdependencies.

- **Dynamic and ephemeral nature**: Cloud-native infrastructure is highly dynamic, with resources scaling up and down based on demand. Additionally, containerized applications and serverless functions have short lifespans, making it difficult to track and monitor assets consistently.

- **Shared responsibility model (SRM)**: In cloud-native environments, the responsibility for security is often shared between the organization and the **cloud service provider** (CSP). Understanding the boundaries of this shared responsibility is crucial to ensure that potential threats are adequately addressed and not overlooked due to misunderstandings.

- **Continuous integration and continuous deployment (CI/CD)**: Cloud-native applications typically rely on CI/CD pipelines for rapid development and deployment. This agility can introduce security risks if not properly managed, as vulnerabilities may be introduced and deployed more quickly.

- **Multi-cloud and hybrid environments**: Many organizations use multiple cloud providers or a mix of cloud and on-premises infrastructure. This diversity adds complexity to threat modeling, as each environment may have different security controls and potential vulnerabilities.

While the cloud-native environment poses unique challenges in conducting threat modeling, it can be successfully executed by performing a series of steps.

Steps for a successful cloud-native threat modeling process

Let us understand each step of threat modeling in a cloud-native environment carefully, as follows:

1. **Define system boundaries**: Start by outlining the components and services that make up your cloud-native environment, including microservices, serverless functions, container orchestrators, databases, and third-party integrations. Create a visual representation, such as a diagram, to illustrate the relationships between components and data flows.

2. **Identify assets and data**: Determine the assets that must be protected within your cloud-native environment, such as sensitive data, critical services, and infrastructure components. Classify these assets based on their importance and the potential impact if compromised, which will help prioritize security efforts and ensure the most critical assets receive the highest level of protection.

3. **Determine threat actors and entry points**: Identify potential threat actors who might target your system, such as external attackers, malicious insiders, or well-intentioned employees making mistakes. Consider their motivations, capabilities, and potential entry points. Entry points could include APIs, user interfaces, or network connections. Understanding the potential avenues for attack will help you establish appropriate security controls to mitigate risk.

4. **Identify and prioritize vulnerabilities**: Analyze your cloud-native environment to identify potential vulnerabilities in your applications, infrastructure, and processes. Use tools such as static and dynamic code analysis, vulnerability scanners, and penetration testing to uncover weaknesses in your environment. Prioritize these vulnerabilities based on the potential impact and the likelihood of exploitation.

5. **Develop mitigation strategies**: For each identified vulnerability, develop appropriate mitigation strategies to reduce the risk. These strategies may include implementing security controls, patching software, or updating configurations. In some cases, accepting the risk or transferring it (for example, through insurance) may be appropriate.

6. **Integrate threat modeling into your CI/CD pipeline**: Integrate threat modeling and security checks into your CI/CD pipeline to ensure that security is considered throughout the development process. Regularly review and update your threat models as your environment and applications evolve to maintain a strong security posture.

7. **Monitor and update your threat models**: Threat modeling is not a one-time exercise; it should be an ongoing process that evolves with your organization and its cloud-native environment. Regularly review and update your threat models to account for changes in your infrastructure, applications, and the threat landscape. This will help you maintain a strong security posture and ensure that your cloud-native environment remains resilient to emerging threats.

8. **Foster collaboration between teams**: Encourage collaboration between development, operations, and security teams to foster a strong security culture within your organization. Ensure that all team members understand the importance of threat modeling and are actively involved in the process. This cross-functional collaboration will help create a more secure environment by encouraging communication, shared understanding, and collective responsibility for security.

9. **Leverage cloud-native security features and tools**: CSPs offer numerous security features and tools designed to help protect cloud-native environments. Take advantage of these offerings to strengthen your security posture. For example, use encryption for data at rest and in transit, implement **identity and access management (IAM)** policies, and configure monitoring and alerting tools to detect potential security incidents.

10. **Stay informed about emerging threats and best practices**: Continuously educate yourself and your team about the latest threats, trends, and best practices in cloud-native security. Participate in industry forums, attend conferences, and follow thought leaders to stay informed about new developments. Apply this knowledge to your threat modeling process to ensure that your cloud-native environment is resilient to emerging threats and aligned with industry best practices.

By following these steps and adapting your threat modeling process to the unique challenges of cloud-native environments, you can proactively identify and address potential threats. This proactive approach will help protect your organization's assets, reduce the likelihood of security incidents, and minimize the potential impact of any breaches.

> **Important note**
>
> One of the other important questions that engineering teams must face which is difficult to answer is: *When to perform threat modeling?* The answer is you should start with threat modeling as soon as the architecture is laid out, and if it is an improvement over an existing architecture, then threat modeling should be performed if there is any architectural or functional change within the system design of the application.

Integrating threat modeling into Agile and DevOps processes

As organizations embrace Agile and DevOps methodologies to accelerate software development and delivery, it becomes increasingly important to integrate threat modeling into these processes. This ensures that security is considered throughout the development life cycle and reduces the likelihood of vulnerabilities being introduced into the final product. In this section, we will explore strategies for incorporating threat modeling into Agile and DevOps processes, discuss the benefits of *shifting security left*, and provide practical guidance for implementation.

Incorporating threat modeling into Agile methodologies

Agile methodologies, such as Scrum and Kanban, prioritize rapid development, iterative improvements, and frequent releases. Integrating threat modeling into Agile processes can be achieved through the following steps:

1. **Include security requirements in user stories**: Embed security requirements directly into user stories, ensuring that they are considered from the beginning of the development process. This includes specifying data protection, authentication, and access control requirements, as well as any other relevant security criteria.

2. **Conduct threat modeling during sprint planning**: Include threat modeling as a task during sprint planning, allowing the team to identify potential risks and vulnerabilities before development begins. This may involve creating a threat model for new features, updating existing threat models to account for changes, or identifying security-related tasks that need to be addressed during the sprint.

3. **Integrate security testing into sprints**: Perform security testing alongside functional testing during each sprint, ensuring that potential vulnerabilities are identified and addressed promptly. This includes static and dynamic code analysis, vulnerability scanning, and penetration testing. By integrating security testing into sprints, the team can quickly identify and remediate any issues that arise during development.

4. **Leverage Agile security champions**: Appoint security champions within the development team who are responsible for promoting security best practices and ensuring that threat modeling and other security activities are carried out during each sprint. These champions should be knowledgeable about security principles and have the authority to advocate for security within the team.

5. **Include security in sprint reviews and retrospectives**: During sprint reviews and retrospectives, discuss security-related achievements and challenges, identify areas for improvement, and adjust the threat modeling process as needed. This continuous improvement mindset will help to ensure that security remains a priority throughout the development life cycle.

Let us now observe how threat modeling differs in DevOps practice in comparison to Agile methodology.

Integrating threat modeling into the DevOps pipeline

DevOps practices aim to bridge the gap between development and operations, enabling organizations to deliver software more quickly and reliably. By integrating threat modeling into the DevOps pipeline, organizations can ensure that security is considered at every stage of the process. Here are some strategies for incorporating threat modeling into the DevOps pipeline:

- **Automate threat modeling tasks**: Leverage automation to simplify and streamline the threat modeling process. This may involve using automated tools to create and update threat models, perform security testing, or assess the security of infrastructure configurations. By automating these tasks, organizations can save time and reduce the risk of human error.

- **Embed security into the CI/CD pipeline**: Integrate threat modeling and security testing into the CI/CD pipeline. This ensures that security checks are performed automatically at each stage of the pipeline, from code commits to deployment. By embedding security into the CI/CD pipeline, organizations can quickly identify and remediate potential vulnerabilities before they reach production environments.

- **Leverage Infrastructure as Code (IaC) for security configuration**: Utilize IaC tools, such as Terraform or CloudFormation, to define and manage the security configurations of your cloud-native environment. By treating IaC, organizations can apply the same threat modeling and security testing processes to infrastructure configurations as they do to application code. This helps to ensure that infrastructure is configured securely and consistently across environments.

- **Monitor and respond to security incidents in real time**: Integrate security monitoring and alerting tools into your DevOps pipeline to detect potential security incidents in real time. Establish a process for responding to security alerts, and ensure that relevant team members are trained to handle incidents efficiently and effectively.

- **Continuously improve security practices**: Regularly review and update your threat modeling process, security testing methodologies, and tooling to align with industry best practices and emerging threats. Continuously evaluate the effectiveness of your security practices and make improvements as needed to maintain a strong security posture.

Let us now seek to understand the different stages where threat modeling can be implemented.

Shifting security left – threat modeling in the early stages of development

"*Shifting security left*" is the practice of incorporating security activities earlier in the development life cycle, rather than treating security as an afterthought or a separate process. By integrating threat modeling into the early stages of development, organizations can proactively identify and address potential vulnerabilities before they become more difficult and costly to remediate. The benefits of shifting security left include the following:

- **Reduced risk**: Early identification and mitigation of vulnerabilities reduce the likelihood of security incidents and minimize the potential impact of any breaches that do occur

- **Faster remediation**: Addressing vulnerabilities during development is generally faster and more efficient than attempting to fix them after deployment, as there is less need to roll back changes or re-architect systems

- **Improved collaboration**: Involving development, operations, and security teams in the threat modeling process from the outset fosters a collaborative security culture and ensures that all team members are aligned in their understanding of security risks and requirements

- **Lower cost**: Identifying and addressing security issues early in the development process can be significantly more cost-effective than dealing with the consequences of a security breach, which may include data loss, regulatory fines, and reputational damage

To successfully shift security left, organizations should consider the following best practices:

- **Educate development teams on security principles**: Ensure that developers understand the importance of security and are familiar with secure coding practices, common vulnerabilities, and potential attack vectors. Providing regular security training and resources can help to build a strong security culture within the development team.

- **Implement security by design**: Incorporate security principles and controls into the design and architecture of your applications and infrastructure from the outset. This may involve using secure design patterns, least privilege principles, and encryption for data at rest and in transit.

- **Adopt a risk-based approach**: Prioritize threat modeling and security activities based on the risk profile of your applications and environment. Focus on addressing the most critical vulnerabilities and assets first to maximize the return on your security investments.

- **Measure and track security metrics**: Establish metrics to measure the effectiveness of your threat modeling and security practices, such as the number of vulnerabilities identified, the time taken to remediate them, and the frequency of security incidents. Continuously track and analyze these metrics to identify areas for improvement and demonstrate the value of your security efforts to stakeholders.

Practical guidance for integrating threat modeling into Agile and DevOps processes

Successfully integrating threat modeling into Agile and DevOps processes requires a combination of cultural, process, and tooling changes. Here are some practical steps to help organizations make this transition:

- **Gain executive buy-in**: Obtaining support from executive leadership is crucial to ensure that security is prioritized within the organization. Present the business case for integrating threat modeling into Agile and DevOps processes, highlighting the potential benefits in terms of risk reduction, cost savings, and regulatory compliance.

- **Build cross-functional teams**: Encourage collaboration between development, operations, and security teams by creating cross-functional teams or embedding security experts within development teams. This will help to break down silos, promote shared understanding, and facilitate more effective communication.

- **Adopt a continuous learning mindset**: Encourage a culture of continuous learning and improvement, with a focus on sharing knowledge, learning from mistakes, and iterating on processes and tooling. This will help to ensure that your threat modeling and security practices remain up to date and aligned with industry best practices.

- **Choose the right tools and technologies**: Select tools and technologies that support your threat modeling and security goals and integrate them into your existing development and operations workflows. This may include threat modeling tools, security testing tools, and monitoring and alerting solutions. Ensure that your chosen tools are compatible with your organization's technology stack and can be easily integrated into your Agile and DevOps processes.

- **Provide training and resources**: Equip your development, operations, and security teams with the knowledge and skills they need to effectively integrate threat modeling into their daily activities. This may involve providing training on secure coding practices, threat modeling methodologies, and the specific tools and technologies your organization has chosen to use.

- **Establish a feedback loop**: Create a feedback loop between development, operations, and security teams to facilitate the sharing of information, lessons learned, and best practices. This can help to identify areas for improvement and ensure that your threat modeling and security practices continue to evolve in line with your organization's needs and the changing threat landscape.

- **Monitor progress and iterate**: Regularly review the effectiveness of your threat modeling and security efforts, using the metrics and indicators you have established. Be prepared to iterate on your processes and tooling as needed and remain open to adopting new practices and technologies as they emerge.

By following these practical steps and embracing a proactive, collaborative approach to threat modeling, organizations can successfully integrate security into their Agile and DevOps processes. This will help to ensure that applications and infrastructure are designed and built with security in mind, reducing the likelihood of vulnerabilities being introduced and minimizing the potential impact of security incidents.

Developing a threat matrix

A threat matrix is a structured framework that helps organizations identify, analyze, and prioritize potential security threats and vulnerabilities within their environment. It serves as a valuable tool in the broader context of threat modeling, enabling teams to systematically assess the various risks they face and develop appropriate mitigation strategies to enhance their overall security posture.

In a cloud-native environment, a threat matrix can be particularly useful, as it allows organizations to navigate the unique security challenges presented by cloud computing, such as multi-tenancy, SRMs, and rapidly evolving infrastructure. By creating a comprehensive threat matrix, organizations can gain a deeper understanding of the different threat actors, attack techniques, and potential vulnerabilities that may impact their cloud-native applications and infrastructure.

The threat matrix can be customized to reflect an organization's specific technology stack, business requirements, and risk tolerance, making it a versatile and adaptable tool for addressing a wide range of security concerns. Furthermore, a well-developed threat matrix can provide valuable insights that inform the design, development, and operation of cloud-native systems, enabling organizations to "shift security left" and proactively address potential issues before they become more difficult and costly to remediate.

In summary, a threat matrix is a critical component of a robust threat modeling process, helping organizations to identify, assess, and prioritize security threats in cloud-native environments. By developing a mindset focused on creating and refining a threat matrix, organizations can more effectively protect their assets, mitigate risks, and maintain a strong security posture in the face of evolving threats and challenges.

Cultivating critical thinking and risk assessment

One of the most important skills for any software engineer to develop is critical thinking; when it comes to threat modeling, this becomes incredibly important. To begin threat modeling, organizations can follow this simple threat model:

Figure 5.1 – Simple threat model mind map

To make this more relatable within the software development life cycle, think about threat modeling as a stage succeeding the system design. Threat modeling is conducted after the architecture and the high-level system design has been decided and laid out. In a nutshell, the goal of conducting such threat modeling exercises is to brainstorm and identify bugs for the team to later fix. The following techniques have worked best for training product teams for such threat modeling exercises.

Fostering a critical thinking mindset

Developing critical thinking skills within your team begins with fostering a mindset that encourages questioning, analysis, and curiosity. Here are some strategies to cultivate critical thinking in your organization:

- **Encourage a culture of curiosity and inquiry**: Create an environment where team members feel comfortable asking questions, challenging assumptions, and exploring alternative perspectives.

Encourage open and honest discussions about potential threats, vulnerabilities, and risks, and promote a collaborative approach to problem-solving.

- **Provide training and resources**: Equip your team members with the knowledge and skills they need to think critically about security threats and risks. This may include providing training on threat modeling methodologies, risk assessment techniques, and secure coding practices, as well as offering resources such as case studies, research papers, and industry reports.

- **Promote diverse perspectives**: Encourage team members from different backgrounds, disciplines, and areas of expertise to contribute their insights and perspectives to the threat modeling process. This diversity of thought can help to uncover potential threats and vulnerabilities that may have been overlooked by a more homogeneous group.

- **Emphasize the importance of reflection and self-assessment**: Encourage team members to regularly reflect on their thought processes, assumptions, and beliefs related to security and risk assessment. This introspection can help individuals identify and challenge cognitive biases that may influence their decision-making.

Developing risk assessment skills

In addition to critical thinking, effective threat modeling requires strong risk assessment skills. To cultivate these skills within your team, consider the following strategies:

- **Adopt a structured approach to risk assessment**: Establish a standardized process for identifying, analyzing, and prioritizing risks, and ensure that all team members are familiar with this approach. This may include using frameworks such as the **National Institute of Standards and Technology (NIST) Cybersecurity Framework** or the **Factor Analysis of Information Risk (FAIR)** model to guide the risk assessment process.

- **Encourage data-driven decision-making**: Encourage team members to base their risk assessments on empirical evidence and data, rather than relying solely on intuition or gut feelings. This may involve collecting data on past security incidents, conducting vulnerability scans, or reviewing industry research and benchmarks.

- **Prioritize risks based on potential impact and likelihood**: Train team members to evaluate risks based on the potential impact on the organization (for example, financial loss, reputational damage, or regulatory penalties) and the likelihood of the risk materializing (for example, the probability of a vulnerability being exploited). This can help to ensure that the most significant risks are addressed first and that resources are allocated effectively.

- **Develop risk communication skills**: Encourage team members to communicate their risk assessments clearly and concisely, using language and terminology that is accessible to non-experts. This can help to ensure that stakeholders understand the potential consequences of identified risks and are able to make informed decisions about risk mitigation strategies.

By cultivating critical thinking and risk assessment skills within your team, you can enhance your organization's ability to identify and address potential threats and vulnerabilities in your cloud-native environment. This proactive approach to security can help to reduce the likelihood of security incidents, minimize the impact of any breaches that do occur, and strengthen your organization's overall security posture.

Threat modeling frameworks

It is important to keep in mind that it is expected for development teams to conduct threat modeling exercises. Other stakeholders for these exercises include product architects, product managers, and application security engineers. However, since these exercises are driven by software engineers, it is important for security engineers to define a framework for the product teams to adhere to. The **Open Worldwide Application Security Project (OWASP)**, Microsoft, and other organizations have developed threat modeling frameworks that are being used across the industry today for people to refer to. We are going to explore some of the frameworks that are widely used, and then define an approach you can use to create a threat modeling approach that suits the business needs of your team.

By following a standardized process, threat modeling frameworks ensure that security assessments are comprehensive, consistent, and aligned with industry best practices. They help organizations create a shared understanding of the security landscape and facilitate collaboration among stakeholders, such as developers, security professionals, and business leaders.

Using threat modeling frameworks is crucial for a variety of reasons. First and foremost, they ensure that organizations consider a wide range of potential threats and vulnerabilities, reducing the likelihood of overlooking critical risks. By providing a structured approach to assessing security risks, organizations can be confident that their analysis is thorough and complete. This comprehensive approach helps to identify security gaps that might otherwise go unnoticed, enabling organizations to address them proactively and effectively.

In addition to their comprehensiveness, threat modeling frameworks promote consistency in security assessments. Adopting a standardized approach allows organizations to maintain consistency in their evaluations, making it easier to compare and track risks across different systems and over time. This consistency is particularly valuable in large organizations or those with complex, interconnected systems, as it ensures that security risks are assessed and managed using the same criteria and processes throughout the organization.

Another significant benefit of threat modeling frameworks is their ability to help organizations prioritize their security efforts. By providing a structured methodology for analyzing and prioritizing risks, threat modeling frameworks enable organizations to allocate their resources more effectively, focusing on the most significant risks first. This risk-based approach to security ensures that organizations can concentrate their efforts on the threats that pose the greatest potential impact, maximizing the **return on investment (ROI)** in security measures.

Collaboration is also a key aspect of threat modeling frameworks. These methodologies often involve multiple stakeholders, fostering collaboration between security, development, and operations teams. This collaboration helps to ensure that all perspectives are considered and leads to more robust and effective security solutions. In the context of cloud-native environments, this collaboration is particularly important, as cloud-native architectures often involve complex interactions between various components, services, and APIs. By working together, security and development teams can identify potential threats and vulnerabilities more effectively and ensure that appropriate security measures are implemented throughout the system.

The application of threat modeling frameworks in the cloud-native landscape is essential to ensure that security is integrated into every aspect of the system from the outset. Cloud-native environments present unique challenges and opportunities for security, with their distributed nature, reliance on microservices, and use of containerization and orchestration technologies. By adopting a systematic approach to threat modeling, organizations can identify and address the specific risks associated with cloud-native architectures, such as data breaches, unauthorized access, and **denial-of-service (DoS)** attacks.

In conclusion, threat modeling frameworks play a vital role in helping organizations identify, analyze, and mitigate security risks in their systems, particularly in the context of cloud-native environments. By providing a comprehensive, consistent, and structured approach to security assessment, these frameworks enable organizations to prioritize their efforts effectively, foster collaboration among stakeholders, and ensure that security measures are integrated throughout the system. As the cloud-native landscape continues to evolve and grow, the use of threat modeling frameworks will become increasingly critical to maintaining the security and resilience of these complex systems.

Now that we have developed a fundamental understanding of threat modeling frameworks, let us discuss a few of the industry-accepted frameworks and how they're applied in the context of cloud-native environments.

STRIDE

The STRIDE threat modeling framework, developed by Microsoft, is a widely adopted approach to identifying and assessing security threats in software systems. STRIDE is an acronym that stands for **Spoofing, Tampering, Repudiation, Information disclosure, DoS, and Elevation of privilege**. These categories represent the primary types of threats that organizations should consider when analyzing the security of their systems and are presented in the following diagram:

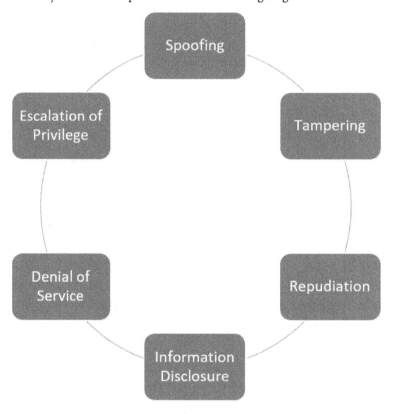

Figure 5.2 – STRIDE threat model framework

Let's go through each of these categories, as follows:

- **Spoofing**: Spoofing refers to impersonating or masquerading as another entity, such as a user, device, or system, in order to gain unauthorized access to sensitive information or perform malicious actions. Examples of spoofing threats include email phishing attacks, IP address spoofing, and identity theft. To mitigate spoofing threats, organizations should implement strong authentication mechanisms, such as **multi-factor authentication** (**MFA**), to ensure that only authorized users can access their systems.

- **Tampering**: Tampering involves unauthorized modification or alteration of data, code, or system components. This can lead to data corruption, unauthorized access, or the execution of malicious code. Examples of tampering threats include **man-in-the-middle (MITM)** attacks, **Structured Query Language (SQL)** injection, and malware infections. To protect against tampering, organizations should implement secure communication channels, validate input data, and ensure the integrity of their software components.

- **Repudiation**: Repudiation refers to the ability of an attacker to deny having performed a malicious action, making it difficult to hold them accountable. This can result from a lack of sufficient logging, auditing, or monitoring capabilities within a system. Examples of repudiation threats include unauthorized transactions or data manipulation without traceability. To address repudiation threats, organizations should implement robust logging and auditing mechanisms to track user activities and provide non-repudiable evidence of malicious actions.

- **Information disclosure**: Information disclosure involves the unauthorized access or exposure of sensitive information, such as personal data, intellectual property, or trade secrets. Examples of information disclosure threats include data breaches, eavesdropping (packet sniffing), and insecure data storage. To mitigate information disclosure threats, organizations should implement strong access controls, data encryption, and secure data storage solutions.

- **DoS**: DoS attacks aim to disrupt the normal operation of a system or service by overwhelming it with malicious traffic or requests, making it unavailable to legitimate users. Examples of DoS threats include **distributed DoS (DDoS)** attacks, resource exhaustion, and application-level DoS attacks. To protect against DoS threats, organizations should implement traffic monitoring, rate limiting, and redundancy measures to ensure the availability and resilience of their systems.

- **Elevation of privilege**: Elevation of privilege refers to the exploitation of vulnerabilities or misconfigurations to gain unauthorized access to resources or perform actions that would normally be restricted. Examples of elevation of privilege threats include privilege escalation attacks, unauthorized access to administrative interfaces, and bypassing security controls. To mitigate elevation of privilege threats, organizations should enforce the **principle of least privilege (PoLP)**, regularly review and update access controls, and patch known vulnerabilities.

The STRIDE framework provides a structured approach to identifying and assessing security threats in software systems by focusing on these six categories. By systematically analyzing each component of a system and determining which of these threat categories apply, organizations can develop appropriate mitigations to address potential vulnerabilities. The STRIDE methodology is particularly useful for organizations looking to integrate security considerations throughout the development life cycle, as it helps to create a shared understanding of potential threats and facilitates collaboration between security, development, and operations teams.

Now, let us try to understand how this threat modeling framework applies to a cloud-native application that is deployed on Kubernetes in **Amazon Web Services** (**AWS**). We will be identifying threat actors, attack vectors, and mitigation techniques for the application while leveraging the STRIDE framework.

We will be considering the architecture of the same To-Do application that we used in *Chapter 2*. Let's first understand the architecture of the application:

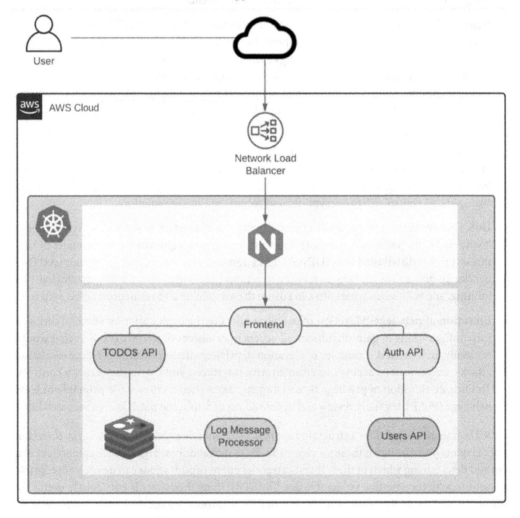

Figure 5.3 – To-Do application deployed in Kubernetes

In this example, we'll create a threat model for a To-Do application hosted on Kubernetes using the STRIDE framework. The application consists of a frontend microservice, an authentication API, a TODOS API, a users API, and a Redis database, all running in separate Pods. We'll use Istio as a service mesh and nginx as an ingress controller. Let's begin by visualizing the application and drawing a **data flow diagram (DFD)**.

DFDs are graphical representations of the flow of data within a system, illustrating how different components, processes, and data stores interact with each other. They are an essential tool for threat modeling, as they help to identify potential security risks and vulnerabilities within the system. By understanding the data flow, security professionals can more effectively prioritize and address potential threats.

The importance of creating a DFD for threat modeling lies in its ability to do the following:

- Provide a clear and concise visualization of the system's architecture, making it easier for team members to understand the system's components and interactions

- Identify trust boundaries within the system, highlighting areas where data transitions from one level of trust to another and where potential security risks might arise

- Pinpoint potential attack surfaces by revealing external entry points and critical internal components that handle sensitive data

- Facilitate communication among team members, allowing for a more effective and collaborative approach to identifying and mitigating security risks

To draw a DFD for threat modeling, follow these steps:

1. **Identify the system's components**: List all the elements involved in the system, such as processes, data stores, and external entities (for example, users and third-party services).

2. **Define the data flows**: Determine how data flows between the components, illustrating the direction and type of data being transmitted.

3. **Identify trust boundaries**: Mark the areas where data moves between components with different trust levels. These boundaries can be physical (for example, between servers) or logical (for example, between applications or services).

4. **Label the diagram**: Provide descriptive labels for components, data flows, and trust boundaries to ensure clarity and facilitate communication among team members.

5. **Review and refine**: Collaborate with team members to review the DFD, identifying any potential security risks or vulnerabilities and refining the diagram as needed to address these concerns.

Creating a DFD is a crucial first step in threat modeling as it lays the foundation for understanding the system's architecture and identifying potential security risks. By visualizing the flow of data within the system, security professionals can better assess potential vulnerabilities and prioritize their efforts to mitigate potential threats.

> **Important note**
>
> You may choose any tool that you're comfortable with to create this diagram. When working collaboratively with a team, remotely, it becomes vital to use a tool whereby everyone in the team can collaborate and work together on creating a blueprint for the diagram, such as Lucidchart or Excalidraw. When working on a threat modeling exercise in person, it is recommended to use a whiteboard, as that makes the process more efficient.

The DFD diagram for the preceding example would look like this:

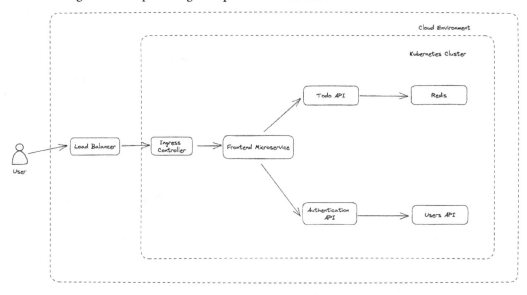

Figure 5.4 – DFD for STRIDE

STRIDE Threat Model

Spoofing

Threat: An attacker impersonating a user to gain unauthorized access to the application or API.

Mitigation:

- Implement strong authentication mechanisms for the frontend web interface and backend APIs using **Open Authorization 2.0 (OAuth 2.0)** with **OpenID Connect (OIDC)**

- Enforce **role-based access control (RBAC)** using Kubernetes' native RBAC feature to manage access to cluster resources
- Configure Istio's **mutual Transport Layer Security (mTLS)** for secure service-to-service communication within the service mesh

Tampering

Threat: An attacker modifying application data, configuration, or code within Pods.

Mitigation:

- Use Kubernetes secrets to store sensitive data such as API keys and credentials securely
- Enable Istio's network policies to limit the communication between Pods only to the required services
- Apply secure coding practices, such as input validation and parameterized queries, to prevent code injection attacks
- Use tools such as **Open Policy Agent (OPA)** Gatekeeper for Kubernetes admission control to limit access to specific objects within your Kubernetes environment, as is also defined in the *OWASP Kubernetes Top Ten* (`https://owasp.org/www-project-kubernetes-top-ten/2022/en/src/K06-broken-authentication`)

Repudiation

Threat: An attacker performing malicious actions without leaving traces that can be attributed to them.

Mitigation:

- Configure logging and monitoring for all application components using a centralized logging system such as Fluentd
- Set up audit logging in Kubernetes to track user activity and changes to cluster resources
- Use tools such as Prometheus and Grafana for monitoring and alerting on suspicious activities

Information disclosure

Threat: An attacker gaining unauthorized access to sensitive data stored in the Redis database or within application components.

Mitigation:

- Enforce strict access controls and authentication for accessing the Redis database and backend APIs
- Encrypt sensitive data using the encryption mechanisms provided by Kubernetes secrets
- Configure Istio to enforce strict egress controls to prevent unauthorized data exfiltration

DoS

Threat: An attacker disrupting the availability of the application by overwhelming the frontend microservice, the authentication API, the TODOS API, or the Redis database with malicious traffic.

Mitigation:

- Use Kubernetes' built-in ReplicaSets and/or StatefulSets features to dynamically scale the application components based on demand

- Set up rate limiting and request throttling using Istio's traffic management features

- Configure nginx IngressController to protect against layer 7 DDoS attacks

Elevation of privilege

Threat: An attacker exploiting vulnerabilities or misconfigurations to gain unauthorized access to resources or perform actions that would normally be restricted.

Mitigation:

- Apply the PoLP to all components and the cluster itself

- Regularly review and update access controls, and patch known vulnerabilities using Kubernetes' rolling updates

- Use a Kubernetes admission controller such as Pod Security Administration or OPA Gatekeeper and network policies to enforce strict security boundaries between different components of the application

By addressing each of the STRIDE categories, we can create a comprehensive threat model for the To-Do application hosted on Kubernetes. This threat model can guide the implementation of appropriate security measures and best practices, ensuring a secure and resilient cloud-native application.

> **Important note**
>
> The goal of conducting threat modeling exercises is to identify bugs before the implementation of the actual feature begins, so it becomes crucial to start thinking about what the actual implementation will look like as soon as the system design for the application begins. It seldom happens that certain architectural changes are required during the threat modeling process to accommodate for security bugs identified, in which case the implementation also changes drastically.

Now that we understand the gist of the STRIDE framework, let's progress to understanding the Process for Attack Simulation and Threat Analysis (PASTA) threat modeling framework, which has been around for a while and has been used extensively in the industry over the past decade.

PASTA

The **PASTA** framework is a risk-centric threat modeling approach that aims to systematically identify and analyze potential threats to a system. Developed by Tony UcedaVélez and Marco M. Morana, PASTA is designed to be comprehensive and iterative, focusing on understanding the system's business context, technical implementation, and potential attack vectors. This framework consists of six main stages, which are outlined as follows:

1. **Define objectives**: The first stage in the PASTA framework is to define the objectives of the threat modeling exercise. This involves understanding the business context and priorities, such as the critical assets, sensitive data, and essential processes that need protection. This step also includes setting the scope, boundaries, and goals for the threat modeling effort.

2. **Define the technical scope**: The second stage is to define the technical scope of the system, which involves creating a detailed understanding of the system's architecture, components, data flows, and interactions. This is typically achieved by developing a system model or DFD that visually represents the system's components, interactions, and trust boundaries. Understanding the technical scope is crucial for identifying potential vulnerabilities and attack surfaces.

3. **Application decomposition**: In the third stage, the system is decomposed into its constituent components and analyzed for potential vulnerabilities. This involves examining the design, implementation, and configuration of each component, as well as understanding the underlying technologies, libraries, and dependencies. By decomposing the application, the threat modeler can identify potential weaknesses and areas where security controls might be lacking.

4. **Threat identification and analysis**: The fourth stage focuses on identifying and analyzing potential threats to the system. Using the information gathered in the previous stages, the threat modeler can systematically enumerate attack vectors, vulnerabilities, and potential threat actors that might target the system. This can be achieved through various techniques, such as brainstorming, risk assessment methodologies, or using **threat intelligence** (TI) feeds. The identified threats are then prioritized based on their likelihood and potential impact.

5. **Attack simulation**: In the fifth stage, identified threats are simulated to understand their potential consequences and validate the effectiveness of existing security controls. This can involve techniques such as penetration testing, vulnerability scanning, or red teaming exercises. By simulating attacks, the threat modeler can gain insights into the system's resilience and identify potential gaps in the security posture.

6. **Risk and impact analysis**: The final stage in the PASTA framework is to analyze the risks and impacts associated with the identified threats. This involves quantifying the potential consequences of each threat, taking into account factors such as the likelihood of exploitation, the severity of the impact, and the effectiveness of existing security controls. Based on this analysis, the threat modeler can develop recommendations for mitigating risks and prioritizing security improvements.

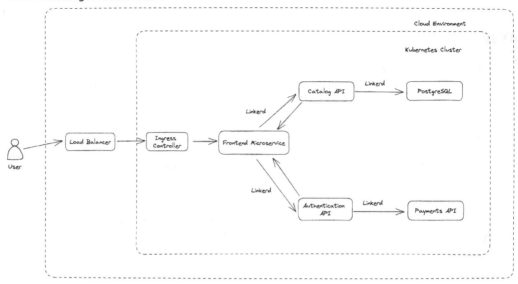

Figure 5.5 – DFD for PASTA

Let's go through the six main stages in the context of this example:

1. **Define objectives**: The primary objectives of the threat modeling exercise are to protect sensitive customer data, maintain the availability of the application, and ensure the integrity of transactions. Key assets include the customer database, payment processing system, and catalog management. The scope includes all components of the application and the underlying Kubernetes infrastructure.

2. **Define technical scope**: The technical scope involves understanding the architecture of the e-commerce application, including the following:

   - Frontend web interface running in a Pod as a microservice

   - RESTful backend API in a separate Pod

   - Catalog microservice managing product listings and inventory in another Pod

   - Payment microservice handling customer payments in yet another Pod

   - PostgreSQL database for storing customer and transaction data running in a dedicated Pod

   - Linkerd service mesh for managing service-to-service communication

   - nginx ingress controller managing external traffic

3. **Application decomposition**: The application is decomposed into its constituent components, listed as follows:

    - **Frontend web interface**: Analyze authentication mechanisms, user input validation, and session management

    - **RESTful backend API**: Review API authentication and authorization, input validation, and data handling

    - **Catalog microservice**: Assess inventory management, data access controls, and potential vulnerabilities in the product listing system

    - **Payment microservice**: Examine payment processing, encryption of sensitive data, and compliance with **Payment Card Industry Data Security Standard (PCI DSS)** requirements

    - **PostgreSQL database**: Evaluate access controls, data encryption, and backup mechanisms

    - **Linkerd service mesh**: Analyze configuration, security policies, and communication between services

    - **nginx ingress controller**: Assess traffic management, load balancing, and security configurations

4. **Threat identification and analysis**: Potential threats, listed as follows, are identified and analyzed:

    - Unauthorized access to customer data, payment information, or application components

    - Injection attacks targeting the backend API, catalog, or payment microservices

    - DoS attacks aimed at disrupting the availability of the application

    - Application security vulnerabilities such as **Cross-Site Scripting (XSS)** and **Cross-Site Request Forgery (CSRF)** can be identified on the frontend web interface

    - Elevation of privilege attacks exploiting misconfigurations or vulnerabilities

5. **Attack simulation**: Simulate the identified threats to understand their potential consequences and validate existing security controls, as follows:

    - Perform penetration testing targeting the frontend web interface, backend API, and microservices

    - Conduct vulnerability scans of the application components and underlying infrastructure

    - Carry out red teaming exercises to emulate real-world attack scenarios

6. **Risk and impact analysis**: Analyze the risks and impacts associated with the identified threats, as follows:

    - Quantify the potential consequences of each threat based on the likelihood of exploitation, the severity of the impact, and the effectiveness of existing security controls

    - Develop recommendations for mitigating risks and prioritizing security improvements

Based on the PASTA threat modeling framework, the e-commerce application can implement appropriate security measures, such as the following:

- Strengthening authentication and authorization mechanisms

- Implementing secure coding practices and input validation

- Encrypting sensitive data both in transit and at rest

- Configuring Linkerd for secure service-to-service communication and enforcing strict network policies

- Regularly updating and patching the application components and Kubernetes infrastructure

By following the PASTA framework, the e-commerce application can achieve a comprehensive understanding of its security posture and prioritize efforts to address identified risks in a cloud-native context.

Now that we have a good grasp of the two major threat modeling frameworks used, we can progress on to the third and final threat modeling framework, which is widely accepted and used in the industry.

LINDDUN

The **Linkability, Identifiability, Non-repudiation, Detectability, Disclosure of information, Unawareness, Non-compliance (LINDDUN)** framework is a privacy-centric threat modeling methodology developed to identify and mitigate privacy threats in information systems. LINDDUN focuses on systematically analyzing the system's data flow and evaluating potential privacy risks that could arise from processing, storing, or sharing personal information. This framework consists of several main steps, which are outlined as follows:

1. **System modeling**: The first step in the LINDDUN framework is to create a model of the system under analysis. This typically involves developing a DFD or another type of visual representation that illustrates the system's components, data flows, trust boundaries, and interactions. The goal is to understand the flow of personal data within the system and identify potential privacy concerns.

2. **Identification of privacy threats**: Using the system model as a basis, the LINDDUN framework provides a set of privacy threat categories, each corresponding to one of the LINDDUN acronym's letters, as set out here:

   - **Linkability**: The ability to link different data elements or sets belonging to the same individual, allowing an attacker to infer additional information about the person

   - **Identifiability**: The risk of identifying an individual based on the data collected or processed by the system, even if the data has been anonymized or pseudonymized

- **Non-repudiation**: The inability of an individual to deny involvement in a transaction or an action, which may lead to privacy risks if sensitive information is disclosed

- **Detectability**: The possibility that an attacker could detect the existence or occurrence of personal data or events within the system, even if they cannot access the actual data

- **Disclosure of information**: The risk of unauthorized access to personal information, either through data breaches, leaks, or improper data-sharing practices

- **Unawareness**: The potential for personal data to be used in ways that were not initially intended or consented to by the individual

- **Non-compliance**: The risk of violating privacy laws, regulations, or policies, which could lead to fines, reputational damage, or other consequences

For each category, privacy threats are systematically identified by analyzing the system model and evaluating how each component, data flow, or interaction could potentially lead to privacy risks.

3. **Privacy risk analysis**: Once privacy threats have been identified, the next step is to analyze the potential risks associated with each threat. This involves evaluating the likelihood of a threat being exploited, the potential impact on individuals' privacy, and the effectiveness of existing privacy controls. Based on this analysis, privacy threats can be prioritized, and appropriate mitigation strategies can be developed.

4. **Privacy threat mitigation**: In this step, mitigation strategies are developed to address the identified privacy threats. The LINDDUN framework provides a set of **privacy-enhancing technologies (PETs)** that can be applied to reduce or eliminate privacy risks. These techniques include data anonymization, encryption, access control, data minimization, and privacy by design principles. By selecting and implementing appropriate PETs, organizations can ensure that their systems are designed and operated with privacy in mind.

5. **Documentation and review**: The final step in the LINDDUN framework is to document the privacy threat analysis, the identified risks, and the mitigation strategies. This documentation serves as a record of the privacy considerations and can be used to demonstrate compliance with privacy laws, regulations, and best practices. Additionally, the LINDDUN framework encourages regular reviews and updates to the threat model as the system and the privacy landscape evolve.

In summary, the LINDDUN threat modeling framework provides a comprehensive and systematic approach to identifying and mitigating privacy risks.

Let's consider an example, and in this example, we'll create a threat model for a telemedicine application hosted on Kubernetes using the LINDDUN framework. The application consists of a frontend web interface, a RESTful backend API, a patient records microservice, a video consultation microservice, and a PostgreSQL database, all running in separate Pods. We'll use Istio as a service mesh and Ambassador as an ingress controller.

We will begin by first creating a DFD based on the description of the architecture laid out:

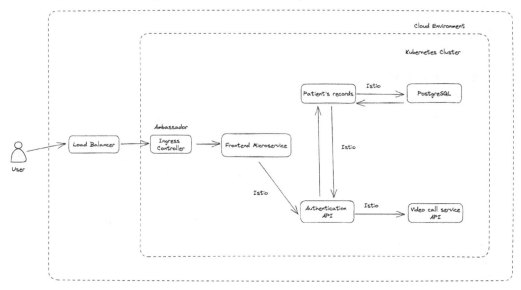

Figure 5.6 – DFD for LINDDUN

Let's go through the five main stages in the context of this example, as follows:

1. **System modeling**: The telemedicine application's architecture includes the following:

    *   Frontend web interface running in a Pod as a microservice

    *   RESTful backend API in a separate Pod

    *   Patient records microservice managing sensitive patient information in another Pod

    *   Video consultation microservice facilitating video calls between patients and doctors in yet another Pod

    *   PostgreSQL database for storing patient records and other sensitive data running in a dedicated Pod

    *   Istio service mesh for managing service-to-service communication

    *   Ambassador ingress controller managing external traffic

2. **Identification of privacy threats**: Using the LINDDUN framework, we identify privacy threats for each component, as follows:

    *   **Linkability**: An attacker could potentially link patient records with video consultation data, revealing sensitive health information

- **Identifiability**: Patient records and video consultation data could be used to identify individuals, breaching their privacy

- **Non-repudiation**: Patients or doctors might not be able to deny involvement in a video consultation, which could have privacy implications

- **Detectability**: An attacker could potentially detect the existence of sensitive patient data or video consultations within the system

- **Disclosure of information**: Unauthorized access to patient records, video consultations, or other sensitive data could occur

- **Unawareness**: Patient data could be used for purposes not initially intended or consented to by the individual

- **Non-compliance**: The application could potentially violate privacy laws or regulations, such as the **Health Insurance Portability and Accountability Act (HIPAA)** or the **General Data Protection Regulation (GDPR)**

3. **Privacy risk analysis**: Analyze the potential risks associated with each identified privacy threat, considering the likelihood of exploitation, the severity of the impact, and the effectiveness of existing privacy controls.

4. **Privacy threat mitigation**: Based on the identified privacy threats, we can apply the following PETs and cloud-native solutions as mitigation strategies:

- **Linkability**: Implement access controls and data segmentation to separate patient records from video consultation data, reducing the ability to link datasets

- **Identifiability**: Anonymize or pseudonymize patient records and video consultation data when not in use to reduce the risk of re-identification

- **Non-repudiation**: Use secure logging mechanisms and data retention policies to balance non-repudiation requirements with privacy concerns

- **Detectability**: Apply network segmentation and strong encryption (for example, mTLS) in the Istio service mesh to reduce the risk of detecting sensitive data

- **Disclosure of information**: Implement RBAC to restrict access to sensitive data and use Kubernetes secrets to store sensitive information securely

- **Unawareness**: Establish strict data usage policies and ensure that data processing activities are transparent and consent driven

- **Non-compliance**: Perform regular privacy audits and assessments to ensure compliance with relevant privacy laws and regulations

5. **Documentation and review**: Document the privacy threat analysis, the identified risks, and the mitigation strategies. Regularly review and update the threat model as the system evolves and the privacy landscape changes.

By following the LINDDUN framework, the telemedicine application can achieve a comprehensive understanding of its privacy posture and prioritize efforts to address identified risks in a cloud-native context.

Remember—no threat modeling framework is perfect, and all threat modeling exercises should be fine-tuned to your organization's tech stack and needs. One of the ways to create your own threat modeling framework that suits your needs is to create a threat matrix, so we are going to learn about one of the most used threat matrices for Kubernetes to understand security attack vectors you should be mindful of.

Kubernetes threat matrix

When learning about the threat modeling process for cloud-native environments, it is important to learn about Kubernetes threats very closely; after all, Kubernetes is the most widely used vendor-agnostic cloud-native platform on the market today. With more and more companies transitioning into using Kubernetes as their cloud-native solution, attackers and security researchers have been trying to find novel attack vectors in compromising apps hosted in Kubernetes and, by extension, compromising the cluster itself.

In this section, we will use Microsoft's Kubernetes threat matrix (`https://microsoft.github.io/Threat-Matrix-for-Kubernetes/`), as presented in the following screenshot, and will analyze the most infamous attack vectors that have been seen in the wild:

Initial Access	Execution	Persistence	Privilege Escalation	Defense Evasion	Credential Access	Discovery	Lateral Movement	Impact
Using Cloud credentials	Exec into container	Backdoor container	Privileged container	Clear container logs	List K8S secrets	Access the K8S API server	Access cloud resources	Data destruction
Compromised images in registry	bash/cmd inside container	Writable hostPath mount	Cluster-admin binding	Delete K8S events	Mount service principal	Access Kubelet API	Container service account	Resource hijacking
Kubeconfig file	New container	Kubernetes CronJob	hostPath mount	Pod / container name similarity	Access container service account	Network mapping	Cluster internal networking	Denial of service
Application vulnerability	Application exploit (RCE)		Access cloud resources	Connect from Proxy server	Applications credentials in configuration files	Access Kubernetes dashboard	Applications credentials in configuration files	
Exposed dashboard	SSH server running inside container					Instance Metadata API	Writable volume mounts on the host	
							Access Kubernetes dashboard	
							Access tiller endpoint	

Figure 5.7 – Microsoft's Kubernetes threat matrix

This threat matrix is inspired by the MITRE ATT&CK framework, a popular knowledge base and methodology for understanding and modeling **advanced persistent threats** (**APTs**). Microsoft's Kubernetes threat matrix focuses on the unique attack vectors and vulnerabilities present in Kubernetes deployments and serves as a valuable reference for organizations looking to improve their Kubernetes security posture. Practically, this threat matrix is used as a discussion point when performing threat modeling exercises; as we understand each attack vector individually, we will discuss and learn how you should use each criterion during your threat modeling exercise.

The matrix is organized into several categories or tactics, each representing a high-level objective that an attacker might pursue. Within each category, the matrix lists specific techniques that an attacker could use to achieve the given objective. For each technique, the matrix provides a brief description, examples, and potential mitigations.

Here are the main categories in Microsoft's Kubernetes threat matrix:

1. **Initial Access**: This category focuses on how an attacker can gain initial access to a Kubernetes cluster, such as through phishing, supply chain compromise, or exploiting public-facing applications.

2. **Execution**: This category deals with the ways an attacker can execute malicious code within the Kubernetes environment, including running malicious containers or using Kubernetes-native tools such as `kubectl`.

3. **Persistence**: Techniques in this category are aimed at maintaining an attacker's foothold in the cluster, even after the initial vulnerability has been patched or the malicious container has been removed. Examples include creating backdoor accounts or exploiting misconfigurations in Kubernetes components.

4. **Privilege Escalation**: This category covers methods an attacker can use to elevate their privileges within the cluster—for example, by exploiting container runtime vulnerabilities or compromising the Kubernetes control plane.

5. **Defense Evasion**: Techniques in this category aim to help an attacker avoid detection by security tools and monitoring systems, such as disabling or tampering with logs, using stealthy network communication, or employing rootless containers.

6. **Credential Access:** This category deals with methods attackers use to steal or obtain credentials for Kubernetes components or applications, such as accessing sensitive data stored in Kubernetes secrets or intercepting API server communication.

7. **Discovery**: Techniques in this category help an attacker gain situational awareness within the Kubernetes environment, such as enumerating cluster nodes, discovering deployed applications, or identifying misconfigurations.

8. **Lateral Movement**: This category covers methods for moving between nodes or containers within the cluster, such as compromising Kubernetes network policies or exploiting container runtime vulnerabilities.

9. **Impact**: Techniques in this category aim to achieve the attacker's ultimate goal, which may include data exfiltration, DoS, or deploying ransomware in the cluster.

On looking at each of the aforementioned points, it is very recognizable that they depict each stage of a pentesting engagement. This also provides a great safety net in conducting threat modeling, to identify and look at each threat individually and progress through other threats, as an attacker would, based on the level of access that they have. Let's closely understand each threat and understand how you would identify and relate threats in your threat modeling engagement.

Initial Access

What access does the attacker have?

At this stage, the attacker is considered to be a user with no access within the Kubernetes cluster, and so it is safe to presume that on looking at threats based on this criterion, for our threat modeling engagement, the attacker is any other user using our platform and trying to gain access to any of the running Pods. Let us examine each attack vector and understand how security engineers think at the time of performing threat modeling for their environment, based on this:

- **Using cloud credentials**: Attackers may attempt to gain unauthorized access to a Kubernetes cluster by obtaining and using cloud provider credentials. These credentials may be leaked accidentally, stolen through phishing attacks, or discovered in exposed repositories. To mitigate this risk during threat modeling, security engineers should ensure the following:

  - Access to cloud provider credentials is restricted to the minimum required personnel.

  - Credentials are securely stored and not hardcoded in application code or configuration files.

  - Regular audits and rotation of credentials are performed.

  - Strong authentication mechanisms, such as MFA, are used.

- **Compromised images in registry**: Attackers might compromise container images stored in a registry by injecting malicious code or using outdated images with known vulnerabilities. During threat modeling, security engineers should consider the following mitigations:

  - Using a private, secured container registry with access controls.

  - Implementing image scanning to identify and address vulnerabilities.

  - Ensuring that images are built from trusted sources and use a minimal base image.

  - Applying strict access controls to the container registry.

  - Using digital signatures to verify the integrity of images before deployment.

- **kubeconfig file**: A `kubeconfig` file contains the necessary information for `kubectl` to connect to a Kubernetes cluster. If an attacker gains access to a kubeconfig file, they may be able to interact with the cluster using the privileges of the file's owner. To address this risk during threat modeling, security engineers should do the following:

  - Restrict access to kubeconfig files using proper file permissions.

  - Avoid storing kubeconfig files in public repositories or insecure locations.

  - Use separate kubeconfig files with different access levels for different users or roles.

  - Monitor and audit the use of kubeconfig files to detect suspicious activity.

- **Application vulnerability**: Vulnerabilities in applications running on the Kubernetes cluster may provide attackers with an initial foothold. To mitigate this risk during threat modeling, security engineers should do the following:

  - Regularly scan application code and dependencies for vulnerabilities.

  - Apply security best practices during application development.

  - Implement robust authentication and authorization mechanisms for applications.

  - Use network segmentation and Kubernetes network policies to limit the attack surface.

- **Exposed dashboard**: An insecurely configured or exposed Kubernetes dashboard can provide attackers with unauthorized access to the cluster. To address this risk during threat modeling, security engineers should do the following:

  - Disable the Kubernetes dashboard if it is not required.

  - Use strong authentication and authorization mechanisms, such as RBAC, for dashboard access.

  - Limit dashboard access to trusted networks or users.

  - Regularly monitor and audit dashboard usage for signs of unauthorized access.

With this, let us look into the next step in a pentest engagement and learn the threat modeling process for it.

Execution

Execution is a critical phase in the attack life cycle, where adversaries seek to run malicious code or commands within the target environment. In the context of Kubernetes, execution can occur in various ways, such as gaining access to a container's shell, exploiting a vulnerable application, or deploying new containers with malicious payloads. Understanding potential execution attack vectors in a Kubernetes cluster is essential for security engineers, as it allows them to identify and mitigate risks before they can be exploited. By incorporating the execution phase into their threat modeling exercises, organizations can proactively secure their Kubernetes deployments and minimize the chances of an attacker gaining unauthorized access or control. Let's examine each attack vector, as follows:

- **Exec into container**: Attackers with access to the Kubernetes cluster might attempt to use the `kubectl exec` or `docker exec` commands to execute arbitrary commands inside running containers. To mitigate this risk during threat modeling, security engineers should do the following:

  - Implement proper RBAC policies to restrict who can use `exec` commands.

  - Monitor and audit the use of `exec` commands within the cluster.

  - Limit the use of privileged containers or containers running as root.

- **Bash/CMD inside container**: Attackers who manage to gain access to a container may execute arbitrary commands via the container's shell (for example, Bash or CMD). To address this risk during threat modeling, security engineers should do the following:

  - Use minimal base images with fewer tools and attack surfaces.

  - Employ read-only filesystems for containers, when possible.

  - Implement proper container isolation using Kubernetes security features such as **PodSecurityPolicies**, **AppArmor**, or **Security-Enhanced Linux (SELinux)**.

- **New containers**: Attackers may attempt to deploy new containers with malicious code or configuration within the cluster. To mitigate this risk during threat modeling, security engineers should do the following:

  - Implement strict RBAC policies to control who can deploy new containers.

  - Use admission controllers (for example, OPA Gatekeeper or Kyverno) to enforce policies on container deployments.

  - Scan container images for vulnerabilities and malicious code before deployment.

  - Monitor and audit container deployments for suspicious activity.

- **Application exploitation**: Attackers may exploit vulnerabilities in applications running within the cluster to gain **remote code execution (RCE)** capabilities. To address this risk during threat modeling, security engineers should do the following:

  - Regularly scan application code and dependencies for known vulnerabilities.

  - Apply security best practices during application development.

  - Implement proper input validation and output encoding to prevent code injection attacks.

  - Limit the attack surface by using network segmentation and Kubernetes network policies.

- **SSH server running inside the container**: Running an SSH server inside a container may provide attackers with a way to execute arbitrary commands or gain unauthorized access. To mitigate this risk during threat modeling, security engineers should do the following:

  - Avoid running SSH servers inside containers unless absolutely necessary.

  - Implement proper access controls for any SSH servers running within containers, such as key-based authentication and IP restrictions.

  - Monitor and audit SSH access to containers.

  - Use ephemeral containers for debugging purposes instead of enabling SSH access.

Persistence

After getting an initial foothold within the cluster, the next step for the attacker would be to create persistent access to the Kubernetes cluster. Let's examine each attack vector in depth and its mitigation strategy:

- **Backdoor container**: Attackers may attempt to create or modify a container to include a backdoor, allowing them to maintain access to the cluster even if the initial point of entry is closed. To mitigate this risk during threat modeling, security engineers should do the following:

  - Implement strict RBAC policies to control who can create or modify containers.

  - Use admission controllers (for example, OPA Gatekeeper or Kyverno) to enforce policies on container deployments.

  - Scan container images for vulnerabilities and backdoors before deployment.

  - Monitor and audit container deployments and modifications for suspicious activity.

- **Writable hostPath mount**: A writable hostPath mount can allow attackers to persist data or malicious code on the underlying Kubernetes node, potentially leading to persistence within the cluster. To address this risk during threat modeling, security engineers should do the following:

  - Limit the use of hostPath mounts in container specifications, especially writable ones.

  - Use PodSecurityPolicies to restrict the use of hostPath mounts.

  - Monitor and audit the use of hostPath mounts for signs of abuse.

  - Regularly scan the underlying nodes for signs of unauthorized access or modifications.

- **Kubernetes CronJob**: Attackers may create or modify Kubernetes CronJobs to maintain persistence by periodically executing malicious code or tasks. To mitigate this risk during threat modeling, security engineers should do the following:

  - Implement proper RBAC policies to control who can create or modify CronJobs.

  - Use admission controllers to enforce policies on CronJob deployments.

- Monitor and audit CronJob creation and modifications for suspicious activity.

- Regularly review and validate the legitimacy of existing CronJobs within the cluster.

Privilege Escalation

Privilege Escalation refers to an attacker attempting to increase their permissions and access within the Kubernetes cluster, often by exploiting certain misconfigurations or vulnerabilities. At this stage, the attacker has initial access and is aiming to elevate their privileges to cause more damage or gain further access. Here are some of the key attack vectors and potential mitigations related to privilege escalation:

- **Privileged Container**: A privileged container is a container that holds nearly the same access as the host machine. An attacker with access to a privileged container can exploit it to gain control over the host machine. During threat modeling, security engineers should consider the following mitigations:

  - Restrict the use of privileged containers as much as possible.

  - Use Kubernetes security contexts to enforce least-privilege access.

  - Regularly audit container privileges and configurations to detect any unnecessary privileges.

  - Use Kubernetes admission controllers such as odSecurityPolicy to prevent the creation of privileged containers.

- **Cluster-admin binding**: The cluster-admin role in Kubernetes has wide-ranging permissions, including the ability to perform any action on any resource. If an attacker gains access to a cluster-admin role, they can control the entire cluster. To mitigate this risk, security engineers should do the following:

  - Use RBAC to limit permissions and enforce least privilege access.

  - Regularly audit RBAC configurations and role bindings.

  - Be careful with service accounts, especially those with cluster-admin bindings.

  - Implement monitoring and alerting for suspicious activity related to the cluster-admin role.

- **hostPath mount**: The hostPath volume mounts a file or directory from the host node's filesystem into a Pod. An attacker who gains access to a Pod with a hostPath mount can potentially read from or write to the host filesystem, leading to a privilege escalation. To prevent this, security engineers should do the following:

  - Limit the use of hostPath volumes to only those cases where it is absolutely necessary.

  - Use Kubernetes admission controllers to prevent the creation of Pods with hostPath volumes.

- Regularly audit Pods for unnecessary hostPath usage.

- Implement monitoring and alerting for suspicious activity related to hostPath volumes.

- **Access Cloud resources**: In a Kubernetes environment hosted on a cloud platform, an attacker may try to escalate privileges by accessing the underlying cloud resources. To mitigate this risk, security engineers should consider the following:

  - Restrict the Kubernetes cluster's permissions to cloud resources using the PoLP.

  - Regularly audit the cluster's cloud permissions.

  - Use cloud provider's IAM roles and policies to limit access.

  - Enable monitoring and logging of cloud resource access, and set up alerts for any unusual activity.

Defense Evasion

Hopefully, on gaining higher privileges, the attacker would want to clear the trace, so as to make it more difficult for the defense tools to detect the attack and either stop it in its tracks or later build the attack chain and implement a fix. Let's look at a few of the defense evasion techniques that attackers use to help them accomplish that:

- **Clear container logs**: Attackers may attempt to clear or tamper with container logs to hide their tracks and avoid detection. To mitigate this risk during threat modeling, security engineers should do the following:

  - Implement centralized logging solutions to collect and store logs outside the cluster (for example, using Elasticsearch, Fluentd, Kibana, or other log management tools).

  - Regularly monitor and analyze logs for signs of suspicious activity or unauthorized access.

  - Use log integrity solutions to detect tampering or log deletion.

  - Set proper access controls and RBAC policies to limit who can access or modify logs.

- **Delete K8s events**: Attackers might try to delete Kubernetes events to cover their tracks and evade detection. To address this risk during threat modeling, security engineers should do the following:

  - Regularly monitor and audit Kubernetes events for signs of unauthorized access or suspicious activity.

  - Implement proper RBAC policies to limit who can delete or modify events.

- Use event export solutions (for example, Kubernetes Event Exporter) to store events outside the cluster for analysis and detection.

- Employ alerting mechanisms to notify administrators of event deletion or unusual patterns.

- **Pod/container name similarity**: Attackers may create Pods or containers with names similar to legitimate ones, making it difficult to identify malicious activity. To mitigate this risk during threat modeling, security engineers should do the following:

  - Implement and enforce naming conventions for Pods and containers to minimize the risk of name collisions.

  - Monitor and audit Pod and container deployments for suspicious activity or naming anomalies.

  - Use admission controllers to enforce policies on Pod and container naming.

  - Regularly review and validate the legitimacy of existing Pods and containers within the cluster.

- **Connect from proxy server**: Attackers may attempt to connect to the Kubernetes cluster through a proxy server to hide their identity and location. To address this risk during threat modeling, security engineers should do the following:

  - Monitor and analyze network traffic for connections from known proxy servers or suspicious IP addresses.

  - Implement network security measures, such as firewalls and security groups, to limit access to the Kubernetes API server and other critical components.

  - Use network segmentation to isolate sensitive components within the cluster.

  - Employ IP-based access controls to restrict access to trusted networks or IP addresses.

Credential Access

This attack criterion is omnipresent throughout the pentest engagement since the attacker will always be on the hunt to find user credentials to elevate their privileges or move laterally within the cluster. Let's look at a few of the attack vectors within this criterion:

- **List K8s secrets**: Attackers may attempt to list Kubernetes secrets in order to access sensitive information, such as credentials or API keys. To mitigate this risk during threat modeling, security engineers should do the following:

  - Implement proper RBAC policies to limit who can list or access secrets.

  - Use admission controllers to enforce policies on secret creation and access.

- Encrypt sensitive data stored in secrets using tools such as Sealed Secrets or Kubernetes External Secrets.

- Regularly monitor and audit secret access for signs of unauthorized activity.

- **Mount service principal**: Attackers might try to mount service principals within a container to gain access to cloud resources or other services. To address this risk during threat modeling, security engineers should do the following:

  - Limit the use of service principals and employ the PoLP.

  - Use PodSecurityPolicies or other security contexts to restrict the mounting of service principals.

  - Monitor and audit the use of service principals within the cluster.

  - Implement proper RBAC policies to control who can create or modify service principals.

- **Access container service account**: Attackers may attempt to access the service account associated with a container to escalate privileges or gain access to sensitive resources. To mitigate this risk during threat modeling, security engineers should do the following:

  - Limit the use of container service accounts and follow the PoLP.

  - Use PodSecurityPolicies or other security contexts to restrict access to service accounts.

  - Regularly monitor and audit service account access and usage for signs of unauthorized activity.

  - Implement proper RBAC policies to control who can create or modify service accounts.

- **Application credentials in configuration files**: Attackers might try to access application credentials stored in configuration files, potentially compromising sensitive data or services. To address this risk during threat modeling, security engineers should do the following:

  - Avoid storing sensitive credentials in plaintext within configuration files.

  - Use Kubernetes secrets or other secret management solutions (for example, HashiCorp Vault) to securely store and manage sensitive credentials.

  - Implement proper access controls for configuration files and limit who can read or modify them.

  - Regularly review and audit the use of credentials within configuration files for signs of unauthorized access or misuse.

Discovery

Similar to credential access, this step of the pentest engagement is omnipresent as well, given that after the preliminary reconnaissance to gain an initial foothold within the cluster, attackers try to keep finding more vulnerabilities inside the cluster. We will try to understand some of the previously discovered and well-known techniques and how you can detect and prevent them within the cluster, as follows:

- **Access the K8s API server**: Attackers may try to access the Kubernetes API server to gather information about the cluster and its resources. To mitigate this risk during threat modeling, security engineers should do the following:

  - Implement strong authentication and authorization mechanisms, such as RBAC and OIDC, to control access to the API server.

  - Use network security measures, such as firewalls and security groups, to limit access to the API server.

  - Monitor and audit API server access for signs of unauthorized activity or suspicious patterns.

  - Employ encryption for data in transit to protect sensitive information exchanged with the API server.

- **Access the Kubelet API**: Attackers might attempt to access the Kubelet API to gather information about nodes and their workloads. To address this risk during threat modeling, security engineers should do the following:

  - Restrict access to the Kubelet API using network security measures, such as firewalls and security groups.

  - Enable and configure Kubelet authentication and authorization to limit who can access the API.

  - Regularly monitor and audit Kubelet API access for signs of unauthorized activity.

  - Use encryption for data in transit to protect sensitive information exchanged with the Kubelet API.

- **Network mapping**: Attackers may perform network mapping to identify cluster resources, communication patterns, and potential attack vectors. To mitigate this risk during threat modeling, security engineers should do the following:

  - Implement network segmentation to isolate sensitive components and limit the potential attack surface.

  - Use network security tools and policies, such as network policies and firewalls, to restrict traffic between Pods and services.

  - Regularly monitor and analyze network traffic for signs of reconnaissance or unauthorized access.

- Employ **intrusion detection and prevention systems (IDPS)** to detect and mitigate potential threats.

- **Access the Kubernetes dashboard**: Attackers might try to access the Kubernetes dashboard to gather information about the cluster and its resources. To address this risk during threat modeling, security engineers should do the following:

  - Limit access to the Kubernetes dashboard using network security measures, such as firewalls and security groups.

  - Implement strong authentication and authorization mechanisms, such as RBAC and OIDC, to control access to the dashboard.

  - Regularly monitor and audit dashboard access for signs of unauthorized activity or suspicious patterns.

  - Consider disabling the dashboard if it is not required for cluster operations.

- **Instance metadata API**: Attackers may attempt to access the instance metadata API to gather information about cloud resources and potentially obtain sensitive data. To mitigate this risk during threat modeling, security engineers should do the following:

  - Restrict access to the instance metadata API using network security measures, such as firewalls and security groups.

  - Limit the exposure of sensitive data through the instance metadata API by configuring the cloud provider's metadata service.

  - Regularly monitor and audit access to the instance metadata API for signs of unauthorized activity.

  - Employ additional security measures, such as workload identity or secret management solutions, to reduce reliance on instance metadata for credential management.

Lateral Movement

An attacker tries to gain as much foothold within the cluster as possible; this is so that the overall impact of the engagement is increased. To make this happen, attackers usually try to perform a lateral movement at every stage possible. Let us look at a few attack vectors and understand how you would want to think about these vectors when performing threat modeling exercises with the operations and the engineering teams:

- **Access cloud resources**: Attackers may attempt to access cloud resources to gather additional information or exploit vulnerabilities. To mitigate this risk during threat modeling, security engineers should do the following:

  - Implement the PoLP to limit access to cloud resources.

- Use network security measures, such as firewalls and security groups, to restrict access to cloud resources.

- Regularly monitor and audit access to cloud resources for signs of unauthorized activity.

- Employ additional security measures, such as workload identity or secret management solutions, to manage access to cloud resources.

- **Container service account**: Attackers might attempt to compromise a container service account to gain access to sensitive resources or escalate privileges. To address this risk during threat modeling, security engineers should do the following:

  - Limit the use of container service accounts and follow the PoLP.

  - Use PodSecurityPolicies or other security contexts to restrict access to service accounts.

  - Regularly monitor and audit service account access and usage for signs of unauthorized activity.

  - Implement proper RBAC policies to control who can create or modify service accounts.

- **Cluster internal networking**: Attackers may try to exploit cluster internal networking to move laterally between services, Pods, or nodes. To mitigate this risk during threat modeling, security engineers should do the following:

  - Implement network segmentation and isolation to restrict communication between sensitive components.

  - Use network policies and firewalls to limit traffic between Pods and services.

  - Regularly monitor and analyze network traffic for signs of unauthorized access or lateral movement.

  - Employ IDPS to detect and mitigate potential threats.

- **Application credentials in configuration files**: Attackers might attempt to access application credentials stored in configuration files to move laterally or gain access to sensitive services. To address this risk during threat modeling, security engineers should do the following:

  - Avoid storing sensitive credentials in plaintext within configuration files.

  - Use Kubernetes secrets or other secret management solutions (for example, HashiCorp Vault) to securely store and manage sensitive credentials.

  - Implement proper access controls for configuration files and limit who can read or modify them.

  - Regularly review and audit the use of credentials within configuration files for signs of unauthorized access or misuse.

- **Writable volume mounts on the host**: Attackers may try to exploit writable volume mounts on the host to move laterally, escalate privileges, or compromise the host system. To mitigate this risk during threat modeling, security engineers should do the following:

  - Limit the use of writable hostPath mounts and follow the PoLP.

  - Use PodSecurityPolicies or other security contexts to restrict the mounting of hostPath volumes.

  - Regularly monitor and audit hostPath volume usage for signs of unauthorized activity.

  - Implement proper RBAC policies to control who can create or modify hostPath volumes.

- **Access the Kubernetes dashboard**: Attackers might attempt to access the Kubernetes dashboard to gather information about the cluster and its resources or move laterally within the cluster. To address this risk during threat modeling, security engineers should do the following:

  - Limit access to the Kubernetes dashboard using network security measures, such as firewalls and security groups.

  - Implement strong authentication and authorization mechanisms, such as RBAC and OIDC, to control access to the dashboard.

  - Regularly monitor and audit dashboard access for signs of unauthorized activity or suspicious patterns.

  - Consider disabling the dashboard if it is not required for cluster operations.

- **Access Tiller endpoint**: Attackers may try to access the Tiller endpoint, which was used in Helm v1 and v2, to compromise the Helm deployment or move laterally within the cluster. To mitigate this risk during threat modeling, security engineers should do the following:

  - Limit access to the Tiller endpoint using network security measures, such as firewalls and security.

  - Update to the latest version of Helm, as it does not use Tiller.

Impact

This step is the most useful at the time of risk assessment and analyzing the attack posture of the application as it encompasses most of the attack vectors we observed previously, but also proves whether there's any true impact of a vulnerability that allows engineers to prioritize bugs identified in terms of the impact of that security bug. Let's have a look at the attack vectors for this criterion:

- **Data destruction**: Attackers may attempt to destroy data stored in the Kubernetes cluster, either to disrupt operations or as a means of retaliation. To address this risk during threat modeling, security engineers should do the following:

  - Implement regular data backups and ensure data is stored redundantly across multiple locations.

  - Use encryption for data at rest and in transit to prevent unauthorized access and tampering.

  - Employ access controls and RBAC policies to limit who can delete or modify data.

  - Monitor and audit data access and manipulation for signs of unauthorized activity or potential data destruction attempts.

- **Resource hijacking**: Attackers might attempt to hijack cluster resources to run malicious workloads, mine cryptocurrencies, or perform other unauthorized activities. To mitigate this risk during threat modeling, security engineers should do the following:

  - Implement resource quotas and limits to prevent excessive resource consumption by any single Pod or user.

  - Use network policies and firewalls to restrict communication between Pods and services, limiting an attacker's ability to move laterally within the cluster.

  - Monitor and analyze cluster resource usage for signs of unauthorized activity or resource hijacking.

  - Employ IDPS to detect and mitigate potential threats.

- **DoS**: Attackers may target a Kubernetes cluster with DoS attacks to disrupt its operation and availability. To address this risk during threat modeling, security engineers should do the following:

  - Implement autoscaling policies to dynamically adjust the number of replicas for a service based on demand, helping to absorb sudden spikes in traffic.

  - Use rate limiting and other application-level security measures to prevent abuse and overloading of APIs or services.

  - Employ a DDoS protection service or a **content delivery network** (**CDN**) to mitigate external attacks.

  - Monitor and analyze network traffic for signs of DoS attacks and respond quickly to mitigate their impact.

Microsoft's Kubernetes threat matrix is an invaluable starting point for security engineers embarking on the threat modeling process for their Kubernetes-based applications. However, it is crucial to recognize that the matrix is not exhaustive and should not be treated as a *one-size-fits-all* solution. After reviewing the threat matrix, security engineers should carefully analyze their unique environment, considering the specific architecture, applications, and configurations deployed within their Kubernetes clusters. This will allow them to identify additional potential attack vectors and vulnerabilities that may not be covered in the matrix. Moreover, security engineers should continuously review and update their threat models, incorporating newly discovered threats, changes in their environment, and evolving best practices in cloud-native security. By doing so, they can ensure that their threat modeling efforts remain relevant and effective in safeguarding their Kubernetes infrastructure against potential attacks.

Summary

This chapter provided a comprehensive overview of threat modeling in the context of cloud-native environments, focusing particularly on Kubernetes. It began with an explanation of threat modeling, its significance in securing cloud-native applications, and the need for organizations to adopt a proactive approach to address potential security risks.

Various threat modeling frameworks, such as STRIDE, PASTA, and LINDDUN, were introduced, along with explanations of their methodologies and use cases. The chapter provided examples of applying these frameworks to Kubernetes-based applications to create threat models, demonstrating how to identify, analyze, and mitigate potential risks in the cloud-native landscape.

Developing a threat matrix was emphasized as an essential part of the threat modeling process, with a particular focus on Microsoft's Kubernetes threat matrix, providing a detailed explanation for each attack vector and guidance on how security engineers can think about and incorporate these factors into their threat modeling exercises.

The chapter also emphasized the importance of cultivating critical thinking and risk assessment skills for threat modeling. Security engineers need to develop a proactive mindset to anticipate potential threats and vulnerabilities in their cloud-native environments and respond accordingly.

In conclusion, this chapter presented a comprehensive guide to threat modeling for cloud-native applications, covering various frameworks, methodologies, and best practices. By understanding and implementing these concepts, security engineers can proactively secure their Kubernetes clusters and mitigate potential risks in the rapidly evolving cloud-native landscape. Now that we have a fair idea on securing the applications, in the next chapter, we will learn about different ways to secure the infrastructure in the cloud environment.

Quiz

Answer the following questions to test your knowledge of this chapter:

- How can organizations effectively integrate threat modeling into their existing Agile and DevOps processes without causing significant disruption or delays in the development cycle? What challenges might they face, and how can they be addressed?

- In the context of cloud-native environments, how can security engineers balance the need for rapid development and deployment with the potential security risks and vulnerabilities that may arise? What strategies or best practices can be adopted to achieve this balance?

- When comparing different threat modeling frameworks such as STRIDE, PASTA, and LINDDUN, what factors should organizations consider when choosing the most suitable framework for their specific cloud-native applications and infrastructure? Can multiple frameworks be used in conjunction or adapted to create a customized approach?

- As cloud-native technologies and the threat landscape continue to evolve, how can security engineers ensure that their threat models remain up to date and relevant? What steps should they take to continuously review and revise their threat modeling processes to address emerging risks and changes in their environment?

- How can organizations foster a culture of security awareness and critical thinking among their development, operations, and security teams to facilitate a proactive approach to threat modeling and risk assessment in cloud-native environments? What challenges might they face in achieving this goal, and how can they be overcome?

Further readings

To learn more about the topics that were covered in this chapter, take a look at the following resources:

- `https://www.microsoft.com/en-us/securityengineering/sdl/threatmodeling`

- `https://www.microsoft.com/en-us/security/blog/2021/03/23/secure-containerized-environments-with-updated-threat-matrix-for-kubernetes/`

- `https://linddun.org/`

- `https://kubernetes.io/docs/concepts/security/`

- `https://owasp.org/www-project-kubernetes-top-ten/`

6

Securing the Infrastructure

In this chapter, we will delve into the critical aspect of securing the infrastructure in cloud-native environments. Infrastructure security is a fundamental component of a comprehensive security strategy as it lays the foundation for protecting applications and data in the cloud. By adhering to best practices and leveraging the right tools and platforms, you can fortify your infrastructure against potential threats and vulnerabilities.

Throughout this chapter, you will gain hands-on experience implementing various security measures for Kubernetes, service mesh, and container security. You will learn how to create and manage network policies, **role-based access control (RBAC)**, and runtime monitoring using tools such as Falco and OPA-Gatekeeper. Additionally, you will explore the security features of Calico as a CNI plugin and understand the importance of container isolation.

We will also discuss the principles of defense in depth and least privilege, guiding you in implementing these principles in a cloud-native environment. You will learn how to design and implement cloud-native platforms using the principle of least privilege, empowering you to manage authentication and authorization effectively.

Finally, by the end of this chapter, you will have a deep understanding of various infrastructure security concepts and will be able to apply these techniques to enhance the security posture of your cloud-native environment.

In this chapter, we're going to cover the following main topics:

- Approach to object access control
- Principles for authentication and authorization
- Defense in depth and least privilege
- Infrastructure components in cloud-native environments

Technical requirements

To follow along with the hands-on tutorials and examples in this chapter, you will need access to the following technologies and installations:

- Kubernetes v1.27

- Calico v3.25

- K9s v0.27.3

- Falco

- OPA Gatekeeper v3.10

Approach to object access control

As we embark on our journey to explore the realm of securing cloud-native infrastructure, one of the primary aspects to consider is the implementation of object access control. In a cloud-native environment, a multitude of components interact with each other, from containers and services to data stores and APIs. To safeguard your infrastructure, it is crucial to establish strict control over who and what can access these objects.

In this section, we will focus on approaches to object access control in the context of Kubernetes, service mesh, and container security. Kubernetes is a widely adopted container orchestration platform that provides a robust framework for deploying, scaling, and managing containerized applications. Service mesh, on the other hand, is an architectural pattern that facilitates secure and efficient communication between microservices. By implementing object access control, you can ensure that only authorized entities interact with your applications, data, and services, significantly mitigating the risk of unauthorized access, data breaches, and other security incidents.

First, we will discuss Kubernetes network policies, which are a native feature of Kubernetes that enables you to define how pods communicate with each other and with other network endpoints. By crafting and managing network policies, you can achieve granular control over the traffic flow in your cluster, allowing or denying specific communication paths as required. Through a hands-on tutorial, you will learn how to create and manage network policies to protect your applications and services.

Then, we will delve into the security features provided by service mesh solutions such as Calico. These platforms offer a range of capabilities, such as **mutual TLS (mTLS)** for encrypting communication between services, access control to restrict traffic flow based on predefined rules, and observability to monitor and analyze the behavior of your applications. In this section, we will provide a hands-on tutorial on implementing security features in Calico to bolster the protection of your microservices.

Kubernetes network policies

As cloud-native applications continue to grow in complexity, ensuring secure and controlled communication between components becomes increasingly important. Kubernetes network policies offer a powerful mechanism to govern network traffic flow within a Kubernetes cluster. They allow you to enforce rules on which components can interact with one another and specify the types of network connections that are permitted or denied. In this section, we will discuss the importance of Kubernetes network policies, the key security considerations when defining them, and walk through an example implementation.

Kubernetes network policies are designed to secure the network layer by defining the allowed or blocked ingress and egress traffic to and from specific pods. By default, all communication between pods in a cluster is allowed, which may not be ideal for a production environment. Network policies provide an effective way to restrict communication between pods, thus limiting the potential attack surface.

When defining network policies, it is crucial to consider the following security aspects:

- **Least privilege principle**: This principle dictates that you should only grant the minimum necessary access required for a specific task. By adhering to this principle, you can effectively limit the attack surface and minimize the risk of unauthorized access or data exfiltration.

- **Segmentation**: By isolating components based on their functionality, you can create a secure environment where each segment is shielded from the others. This can prevent the lateral movement of an attacker within the cluster.

- **Deny by default**: By default, no network policies are enforced in a Kubernetes cluster, and all communication between pods is allowed. To create a more secure environment, it is recommended to deny all traffic by default and only explicitly allow the necessary connections.

- **Monitoring and logging**: It is essential to monitor and log network traffic to detect any anomalies or malicious activity. Continuous monitoring enables you to proactively identify and respond to potential security threats.

Now, let's discuss an example of implementing a network policy. Suppose you have a three-tier application consisting of a frontend, a backend, and a database. You want to restrict communication such that only the frontend can access the backend, and only the backend can access the database. Here's a sample network policy to enforce these restrictions:

```
kind: NetworkPolicy
apiVersion: networking.k8s.io/v1
metadata:
  name: frontend-to-backend
```

```
spec:
  podSelector:
    matchLabels:
      app: backend
  policyTypes:
  - Ingress
  ingress:
  - from:
    - podSelector:
        matchLabels:
          app: frontend
    ports:
    - protocol: TCP
      port: 80
---
kind: NetworkPolicy
apiVersion: networking.k8s.io/v1
metadata:
  name: backend-to-database
spec:
  podSelector:
    matchLabels:
      app: database
  policyTypes:
  - Ingress
  ingress:
  - from:
    - podSelector:
        matchLabels:
          app: backend
    ports:
    - protocol: TCP
      port: 3306
```

This configuration creates two network policies: one that allows traffic from the frontend to the backend on port 80 (HTTP) and another that permits traffic from the backend to the database on port 3306 (MySQL).

Different scenarios may require you to implement network policies in different ways. For instance, you might want to restrict access to specific namespaces, allow communication only from specific IP ranges, or require network traffic to be encrypted using TLS.

In cloud-native environments, numerous components interact with each other, including pods, services, and data stores. Object access control refers to the mechanisms and strategies used to ensure that only authorized entities can access these objects in the system.

Kubernetes network policies are one such mechanism that can help you establish and enforce access control rules at the network level. By defining these policies, you can control the communication between pods and other network endpoints, ensuring that only authorized connections are allowed. This way, you can protect sensitive application components and data stores from unauthorized access and potential security threats.

In summary, Kubernetes network policies play a vital role in the broader topic of object access control by providing a means to secure and control network traffic flow between the various components in a cloud-native environment. By implementing appropriate network policies, you can enhance the overall security posture of your infrastructure and ensure that only authorized entities can interact with your applications, data, and services.

To enforce the preceding custom network policy, save the file as `networkPolicy.yaml` and run the following command inside your Kubernetes cluster:

```
$ kubectl apply -f networkPolicy.yaml
```

> **Important note**
>
> On implementing the Kubernetes network policy inside the cluster, it is important to remember that no other network connection would be allowed besides the ingress/egress policies defined in the YAML definition. For Kubernetes internal protocols such as DNS resolution, a global network policy should be defined to allow connections pertaining to that protocol, such as DNS.

While Kubernetes network policies offer a useful way to control communication between components in a cluster, they might not be sufficient for addressing all security and networking requirements. Calico is an open source networking and network security solution for containers, **virtual machines** (**VMs**), and native host-based workloads that provide additional functionality and enhanced security features. It is often used alongside Kubernetes to extend the capabilities of network policies and better secure the cluster.

Calico

Calico is designed to simplify, scale, and secure cloud-native environments by providing advanced networking capabilities and powerful security features. It leverages a pure Layer 3 approach to networking, which allows for greater scalability and performance compared to traditional Layer 2 solutions. Some of the key features of Calico include the following:

- **Network policy enforcement**: Calico can enforce Kubernetes network policies and its own network policies, providing fine-grained control over communication between components in a cluster.

- **Egress access controls**: While Kubernetes network policies primarily focus on ingress traffic, Calico extends this functionality by offering egress access controls, enabling you to better secure outbound traffic from your workloads.

- **Security group integration**: Calico integrates with public cloud security groups, allowing you to apply consistent security policies across hybrid and multi-cloud environments.

- **IP address management**: Calico provides advanced IP address management features, such as IP pool assignment and dynamic allocation, making it easier to manage and maintain your cluster's IP addresses.

Using Calico with Kubernetes

To use Calico with Kubernetes, you need to install and configure it on your cluster. You can follow the official Calico documentation to get started: `https://docs.projectcalico.org/getting-started/kubernetes/`.

Once Calico has been installed and configured, you can leverage its features to create and manage network policies that extend and enhance the capabilities of Kubernetes native network policies.

The following reasons should be factored in when you're considering using Calico as the network policy-enforcing agent:

- **Enhanced Security**: Calico offers advanced features such as egress access controls, security group integration, and additional policy enforcement options that provide a more comprehensive security solution compared to basic Kubernetes network policies.

- **Scalability**: Calico's Layer 3 networking approach allows for better scalability, making it suitable for large-scale deployments and high-performance workloads.

- **Multi-cloud support**: Calico supports various platforms, including public clouds, private clouds, and on-premises environments. This enables security engineers to apply consistent security policies across different infrastructure setups.

- **Ease of management**: Calico provides advanced IP address management and policy enforcement capabilities, simplifying the process of managing and maintaining network security in a Kubernetes cluster.

In conclusion, Calico is an essential tool for enhancing the security and networking capabilities of a Kubernetes cluster. By using Calico, security engineers can implement more robust and scalable network policies that better protect their cloud-native infrastructure. With its advanced features and multi-cloud support, Calico is an excellent choice for organizations looking to strengthen their network security posture in a Kubernetes environment.

Next, we will learn how to implement Calico network policies for an application deployed in a Kubernetes cluster. We will cover a few use cases to showcase the versatility and power of Calico in achieving object access control inside the Kubernetes cluster. The following are the prerequisites that you need to have installed to successfully run the tutorials provided next:

- A running Kubernetes cluster with Calico installed and configured. You can follow the official Calico documentation to set up Calico: `https://docs.projectcalico.org/getting-started/kubernetes/`.

- `kubectl` installed and configured to interact with your Kubernetes cluster.

Use case 1 – isolating a namespace

In this use case, we will create a Calico network policy to isolate a specific namespace, allowing only ingress traffic from another namespace. You may perform the following steps to enforce this network policy:

1. Create two namespaces, `namespace-a` and `namespace-b`:

   ```
   $ kubectl create namespace namespace-a
   $ kubectl create namespace namespace-b
   ```

2. Deploy a sample application in both namespaces:

   ```
   $ kubectl run nginx --image=nginx:1.14.2 --port=80 -n
   namespace-a
   $ kubectl run nginx --image=nginx:1.14.2 --port=80 -n
   namespace-b
   ```

3. Create a Calico network policy to allow ingress traffic from `namespace-a` to `namespace-b` and save the following YAML in a file called `isolate-namespace.yaml`:

   ```
   apiVersion: projectcalico.org/v3
   kind: GlobalNetworkPolicy
   metadata:
     name: namespace-a-to-namespace-b
   spec:
     selector: projectcalico.org/namespace == 'namespace-b'
     ingress:
       - action: Allow
         source:
           selector: projectcalico.org/namespace == 'namespace-a'
     types:
       - Ingress
   ```

4. Apply the Calico network policy:

```
$ calicoctl apply -f isolate-namespace.yaml
```

This policy isolates namespace-b by only allowing ingress traffic from namespace-a. It demonstrates how Calico can be used to achieve object access control by isolating namespaces inside the Kubernetes cluster.

Use case 2 – restricting egress traffic to a specific external IP range

In this use case, we will create a Calico network policy to restrict egress traffic from our application to a specific external IP range:

1. Deploy a sample application in the default namespace:

```
$ kubectl run nginx --image=nginx:1.14.2 --port=80
```

2. Create a Calico network policy to restrict egress traffic to the 192.0.2.0/24 IP range and save the following YAML in a file named restrict-egress.yaml:

```
apiVersion: projectcalico.org/v3
kind: GlobalNetworkPolicy
metadata:
  name: restrict-egress
spec:
  selector: projectcalico.org/namespace == 'default'
  egress:
    - action: Allow
      destination:
        nets:
          - 192.0.2.0/24
  types:
    - Egress
```

3. Apply the Calico network policy:

```
$ calicoctl apply -f restrict-egress.yaml
```

This policy restricts egress traffic from the default namespace to the 192.0.2.0/24 IP range. It demonstrates how Calico can be used to achieve object access control by controlling egress traffic from the Kubernetes cluster.

Use case 3 – restricting ingress traffic based on port and protocol

In this use case, we will create a Calico network policy to restrict ingress traffic to our application based on port and protocol:

1. Deploy a sample application in the `default` namespace:

   ```
   $ kubectl run nginx --image=nginx:1.14.2 --port=80
   ```

2. Expose the application using a `NodePort` service:

   ```
   $ kubectl expose pod/nginx --type=NodePort --port=80
   ```

3. Create a Calico network policy to restrict ingress traffic to the application based on port 80 and the TCP protocol and save the following YAML in a file named `restrict-ingress-port-protocol.yaml`:

   ```
   apiVersion: projectcalico.org/v3
   kind: GlobalNetworkPolicy
   metadata:
     name: restrict-ingress-port-protocol
   spec:
     selector: projectcalico.org/namespace == 'default'
     ingress:
       - action: Allow
         protocol: TCP
         source:
           nets:
             - <your-source-ip-range>
         destination:
           ports:
             - 80
     types:
       - Ingress
   ```

   Replace `<your-source-ip-range>` with the IP subnet range that you want to allow the ingress traffic on.

4. Apply the Calico network policy:

   ```
   $ calicoctl apply -f restrict-ingress-port-protocol.yaml
   ```

This policy restricts ingress traffic to the application based on port 80 and the TCP protocol. It demonstrates how Calico can be used to achieve object access control by controlling ingress traffic based on specific ports and protocols.

In conclusion, this tutorial has shown you how to implement Calico network policies for a Kubernetes application, with three use cases highlighting the flexibility and power of Calico in achieving object access control inside a Kubernetes cluster.

Principles for authentication and authorization

Authentication and authorization are essential security concepts that govern access control in distributed systems such as Kubernetes. Both processes play a crucial role in protecting sensitive data and resources from unauthorized access. In this section, we will explain the fundamentals of authentication and authorization, focusing on their importance and implications in a Kubernetes cluster.

First, let's understand and analyze each term individually and then seek to learn what they mean and their significance in cloud-native environments.

Authentication

Authentication is the process of verifying the identity of a user, client, or system attempting to access a resource. In the context of Kubernetes, authentication primarily involves confirming the identity of users or service accounts trying to access the Kubernetes API. Kubernetes supports various authentication mechanisms, including client certificates, bearer tokens, and external authentication providers such as **OpenID Connect** (**OIDC**) and LDAP.

When a request reaches the Kubernetes API server, it first goes through an authentication process. The API server checks the request's credentials (for example, client certificate or token) and validates them against the configured authentication mechanisms. If the credentials are valid, the request proceeds to the next stage, which is authorization. If the credentials are invalid or missing, the request is denied, and an error message is returned to the user.

Authorization

Authorization is the process of determining whether an authenticated user or service account has permission to perform an action on a specific resource. Kubernetes implements authorization through RBAC, which allows you to define fine-grained permissions based on roles and role bindings.

Roles define a set of permissions, such as the ability to create, update, delete, or list specific resources. Role bindings associate these roles with users, groups, or service accounts, granting them the defined permissions. Kubernetes supports two types of roles: **ClusterRoles** and **Roles**. ClusterRoles apply to the entire cluster, while Roles are scoped to specific namespaces.

After successful authentication, the Kubernetes API server evaluates the request against the configured authorization policies. If the authenticated user has the necessary permissions to perform the requested action, the request is allowed to proceed. Otherwise, the request is denied, and an error message is returned to the user.

In addition to RBAC, Kubernetes supports other authorization modules such as **attribute-based access control (ABAC)**, Node, and Webhook. However, RBAC is the most widely used and recommended approach due to its flexibility and ease of management.

Importance of authentication and authorization

Implementing robust authentication and authorization processes is essential for ensuring the security and integrity of your Kubernetes cluster. By properly configuring and managing access control, you can prevent unauthorized users and malicious actors from gaining access to sensitive resources or disrupting the cluster's operation.

Adopting strong authentication and authorization practices also helps organizations meet regulatory and compliance requirements, as well as maintain trust with their customers and partners. In the context of Kubernetes, this involves configuring and enforcing access control policies, regularly reviewing and updating permissions, and monitoring and auditing access to the cluster's resources.

In summary, authentication and authorization are vital components of a comprehensive security strategy in any cloud environment. By understanding and effectively implementing these processes, you can significantly enhance the security posture of your cluster and protect your organization's valuable data and resources.

Kubernetes authentication and authorization mechanisms

Continuing from our previous discussion on authentication and authorization, let's explore some cloud-native tools and built-in Kubernetes features that can help you implement access control in your cluster. These tools and mechanisms enable you to manage access to your cluster's resources effectively and securely.

Kubernetes native authentication and authorization

Kubernetes provides native support for various authentication and authorization mechanisms. Let's discuss them in more detail.

Authentication mechanisms

Some of the ways to facilitate authentication in Kubernetes natively are as follows:

- **X.509 client certificates**: Kubernetes can use client certificates to authenticate users:

  - *How it works*: The Kubernetes API server is configured to accept client certificates issued by a trusted **certificate authority (CA)**. Users present their client certificates when making API requests, and the API server verifies them against the trusted CA.

- *Usage*: To use client certificates, generate and sign them using a trusted CA, configure the API server to trust the CA, and then distribute the certificates to users.

- *Justification*: X.509 client certificates provide a secure way of authenticating users, making them suitable for production environments. However, managing certificates can be more complex compared to other authentication methods.

- **Static token file**:

  - *How it works*: Kubernetes uses a file containing bearer tokens to authenticate users. Each token maps to a username, user ID, and a set of groups.

  - *Usage*: To use static tokens, create a file with token entries, configure the API server to use the token file, and distribute tokens to users.

  - *Justification*: Static tokens are less secure and recommended only for testing or non-production environments due to the potential for token leakage and limited scalability.

- **Static password file**: Similar to static tokens, Kubernetes can use a file containing username and password pairs for authentication:

  - *How it works*: Kubernetes uses a file containing username and password pairs for authentication

  - *Usage*: To use static passwords, create a file with username and password entries, configure the API server to use the password file, and distribute passwords to users

  - *Justification*: Like static tokens, static passwords are less secure and recommended for testing or non-production environments only

- **Service account tokens**:

  - *How it works*: Service account tokens are automatically generated for service accounts within the cluster and can be used to authenticate API requests

  - *Usage*: To use service account tokens, create a service account in Kubernetes, and use its token for authentication when making API requests

  - *Justification*: Service account tokens are suitable for automating tasks and managing resources within the cluster but should not be used for human users

- **OIDC tokens**:

  - *How it works*: Kubernetes can authenticate users with OIDC identity providers, enabling **single sign-on (SSO)** across multiple applications

  - *Usage*: To use OIDC tokens, configure the API server to trust an OIDC provider, and set up an OIDC client for Kubernetes

- *Justification*: OIDC tokens provide a secure and convenient way of authenticating users in a centralized manner, making them suitable for production environments

- **Webhook token authentication**:

  - *How it works*: Kubernetes delegates authentication to an external webhook service that returns the user's identity based on the provided token

  - *Usage*: To use webhook token authentication, configure the API server to use a webhook service, and set up the webhook service to authenticate tokens

  - *Justification*: Webhook token authentication allows you to integrate custom authentication services or leverage existing authentication infrastructure, providing flexibility and security

Authorization mechanisms

Here are some of the ways you can facilitate authorization in Kubernetes natively:

- **RBAC**:

  - *How it works*: RBAC allows you to create fine-grained permissions based on roles and role bindings, which are associated with users, groups, or service accounts.

  - *Usage*: To use RBAC, create roles and role bindings in Kubernetes, and assign them to users, groups, or service accounts.

  - *Justification*: RBAC is the recommended authorization method in Kubernetes due to its flexibility, granularity, and ease of management. It allows you to define permissions based on the principle of least privilege, ensuring that users and applications have only the necessary permissions to perform their tasks.

- **ABAC**:

  - *How it works*: ABAC uses policies to grant or deny access based on attributes such as user, resource, and environment.

  - *Usage*: To use ABAC, configure the API server to use ABAC policies, and create policy files containing the necessary rules.

  - *Justification*: While ABAC can provide fine-grained control, it can be more challenging to manage and maintain compared to RBAC. Due to its complexity, ABAC is not recommended for new deployments and should only be used in existing environments where migration to RBAC is not feasible.

- **Node authorization**:

  - *How it works*: The Node Authorizer restricts the access of kubelet nodes to the resources they manage, preventing unauthorized access to sensitive resources

  - *Usage*: To use the Node Authorizer, configure the API server and kubelet to use the Node Authorizer and NodeRestriction admission plugins, respectively

  - *Justification*: The Node Authorizer is essential for securing communication between the API server and kubelet nodes, ensuring that nodes can only access the resources they are responsible for managing

- **Webhook authorization**:

  - *How it works*: Webhook authorization delegates authorization decisions to an external webhook service that returns the decision based on the provided request

  - *Usage*: To use webhook authorization, configure the API server to use a webhook service, and set up the webhook service to evaluate authorization requests

  - *Justification*: Webhook authorization allows you to integrate custom authorization services or leverage existing authorization infrastructure, providing flexibility and security

Authentication and authorization best practices

When provisioning authentication and authorization policies within Kubernetes, organizations are expected to follow the following best practices:

- Always follow the principle of least privilege when granting access to resources

- Regularly review and update your access control policies to ensure they remain effective and relevant

- Monitor and audit access to your cluster's resources to detect potential security issues and unauthorized access attempts

- Make sure you secure and protect the management of authentication credentials, tokens, and certificates to avoid unauthorized access

- Use a centralized and secure secrets management solution, such as HashiCorp Vault or Kubernetes Secrets, to store sensitive information such as credentials and API keys

- When using Webhook token authentication or webhook authorization, ensure that the webhook service is secured, and communication is encrypted using TLS

- Consider using network policies to restrict traffic between pods and namespaces, adding another layer of security

- Keep your cluster and its components up to date with the latest security patches and updates to minimize potential vulnerabilities

- Implement regular security audits and vulnerability scanning to identify and remediate any weaknesses in your cluster's security posture

In summary, understanding and effectively using Kubernetes' native features and additional cloud-native tools for authentication and authorization can help you create a secure and robust cluster environment. By adhering to best practices and using the appropriate tools, you can ensure that your infrastructure remains secure and compliant with your organization's policies. Regular monitoring and updates will also help you maintain a strong security posture and protect your valuable resources from potential threats.

After learning about the best practices for enforcing authentication and authorization, let's now look into learning about an industry-wide tool for enforcing security and compliance policies for admission control in Kubernetes.

OPA Gatekeeper

Open Policy Agent (**OPA**) Gatekeeper is an open source policy management system specifically designed for Kubernetes environments. It is built on top of the OPA and provides a way to enforce flexible and fine-grained policies across a Kubernetes cluster. OPA Gatekeeper is a Kubernetes-native solution that uses the Kubernetes admission control system to validate, mutate, or reject API requests based on policies defined by users. Some of the key features of using Gatekeeper are as follows:

- **Policy as Code**: OPA Gatekeeper allows you to define policies using the Rego language, a high-level, declarative language for expressing policies

- **Dynamic policy enforcement**: OPA Gatekeeper can enforce policies dynamically as resources are created or updated, preventing policy violations

- **Extensibility**: OPA Gatekeeper supports custom logic for defining policies, allowing you to create policies tailored to your organization's specific requirements

- **Audit and reporting**: OPA Gatekeeper can generate audit reports to help you identify policy violations and maintain compliance

- **Integration**: OPA Gatekeeper can be easily integrated with various Kubernetes resources, such as **custom resource definitions** (**CRDs**)

Based on these key features, let's learn about some of the use cases for leveraging these features so that we can provision admission controls in a Kubernetes environment.

Use cases for OPA Gatekeeper

Later in this chapter, we will learn how to implement Gatekeeper policies for applying security protections and putting compliance checks in place. Some of the significant use cases of Gatekeeper are as follows:

- **Security and compliance**: OPA Gatekeeper helps organizations maintain security and compliance by enforcing policies that restrict the creation of insecure resources, limit access to sensitive data, and ensure the proper configuration of cluster components

- **Resource management**: OPA Gatekeeper allows you to define policies that control resource allocation and usage, such as limiting the number of replicas for a deployment or ensuring that all pods have resource limits and requests defined

- **Namespace isolation**: OPA Gatekeeper can enforce policies that restrict cross-namespace communication, ensuring that namespaces are logically isolated from each other

- **Label and annotation management**: OPA Gatekeeper can enforce policies that require specific labels or annotations on resources, ensuring consistent metadata across the cluster

- **Network policy enforcement**: OPA Gatekeeper can be used to enforce network policies that control ingress and egress traffic within the cluster, improving network security and segmentation

Now that we understand both the Kubernetes native authentication mechanism and OPA Gatekeeper, let's do a quick comparison to understand why you might want to choose one over the other.

OPA Gatekeeper versus Kubernetes native admission controllers

Kubernetes native admission controllers are built-in components that intercept, validate, and potentially modify API requests based on predefined rules. These controllers can be useful for basic policy enforcement but have certain limitations compared to OPA Gatekeeper. They are as follows:

- **Flexibility**: Kubernetes native admission controllers often have limited flexibility when it comes to defining custom policies. OPA Gatekeeper, on the other hand, allows you to create highly customizable policies using the Rego language.

- **Policy management**: OPA Gatekeeper offers a more centralized and streamlined policy management system compared to Kubernetes native admission controllers. With OPA Gatekeeper, policies are defined, audited, and enforced consistently across the cluster.

- **Extensibility**: Kubernetes native admission controllers generally have a fixed set of functionality, whereas OPA Gatekeeper can be extended with custom logic and integrated with other tools and platforms.

- **Audit and reporting**: OPA Gatekeeper provides built-in support for auditing and reporting, which is not available in Kubernetes native admission controllers. This feature allows organizations to maintain compliance and quickly identify policy violations.

Based on these limitations of Kubernetes' native solutions, organizations might be more inclined toward leveraging Gatekeeper.

Why organizations should consider OPA Gatekeeper

Organizations should consider using OPA Gatekeeper for the following reasons:

- **Enhanced security**: OPA Gatekeeper provides more fine-grained and flexible policy enforcement capabilities compared to Kubernetes native admission controllers, resulting in a more secure cluster environment

- **Simplified policy management**: OPA Gatekeeper offers a unified, easy-to-use system for managing policies across your entire Kubernetes cluster, simplifying policy management and maintenance

- **Compliance**: With built-in auditing and reporting capabilities, OPA Gatekeeper enables organizations to maintain compliance with internal and external regulations, providing insights into policy violations and helping to remediate any issues

- **Customizability and extensibility**: OPA Gatekeeper's support for custom logic and integration with other tools and platforms make it a versatile solution that can be adapted to meet the specific needs and requirements of different organizations

- **Efficient resource utilization**: By enforcing policies that control resource allocation and usage, OPA Gatekeeper helps organizations optimize the use of their resources and avoid resource wastage

In conclusion, OPA Gatekeeper provides a powerful, flexible, and extensible solution for policy enforcement in Kubernetes environments. It addresses the limitations of Kubernetes native admission controllers and offers enhanced security, simplified policy management, and better compliance capabilities. By leveraging OPA Gatekeeper, organizations can create a more secure and well-governed Kubernetes cluster that aligns with their unique requirements and best practices.

Implementing Gatekeeper security policies

Now, let's learn how to use Gatekeeper as a tool to implement policies across your cluster environment and improve the security posture for your cloud environment. Before reading up and following this tutorial, ensure that OPA Gatekeeper is installed in your Kubernetes cluster. Follow the official installation guide at `https://github.com/open-policy-agent/gatekeeper#installation-instructions`:

- **Enforcing pod resource limits and requests with OPA Gatekeeper**:

  Enforcing pod resource limits and requests is critical for ensuring efficient resource utilization and preventing resource starvation within a Kubernetes cluster. By setting appropriate limits and requests, you can prevent a single application from consuming excessive resources and negatively impacting other applications running on the same node. OPA Gatekeeper makes it easy to enforce these constraints across your cluster.

I. **Create a constraint template**: Create a file named `pod-resource-limits-template.yaml` with the following content:

```yaml
apiVersion: templates.gatekeeper.sh/v1beta1
kind: ConstraintTemplate
metadata:
  name: k8spodresourcelimits
spec:
  crd:
    spec:
      names:
        kind: K8sPodResourceLimits
        listKind: K8sPodResourceLimitsList
        plural: k8spodresourcelimits
        singular: k8spodresourcelimits
  targets:
    - target: admission.k8s.gatekeeper.sh
      rego: |
        package k8spodresourcelimits

        violation[{"msg": msg}] {
            input.review.object.kind == "Pod"
            not input.review.object.spec.containers[_].resources.limits.cpu
            not input.review.object.spec.containers[_].resources.limits.memory
            not input.review.object.spec.containers[_].resources.requests.cpu
            not input.review.object.spec.containers[_].resources.requests.memory
            msg := "Pod must have CPU and memory limits and requests set for all containers"
        }
```

II. Apply the constraint template to your cluster by running the following command:

```
$ kubectl apply -f pod-resource-limits-template.yaml
```

III. **Create a constraint**: Now, create a file named `pod-resource-limits-constraint.yaml` with the following content:

```yaml
apiVersion: constraints.gatekeeper.sh/v1beta1
kind: K8sPodResourceLimits
metadata:
  name: pod-resource-limits
```

```
spec:
  match:
    kinds:
      - apiGroups: [""]
        kinds: ["Pod"]
```

IV.   Apply the constraint to your cluster:

```
$ kubectl apply -f pod-resource-limits-constraint.yaml
```

V.   **Test the constraint**: Create a file named `test-pod.yaml` with the following content:

```
apiVersion: v1
kind: Pod
metadata:
  name: test-pod
spec:
  containers:
  - name: test-container
    image: nginx
```

VI.   Try to apply the test pod using the following command:

```
$ kubectl apply -f test-pod.yaml
```

VII.   You should receive an error message indicating that the pod must have CPU and memory limits and requests set for all containers. Update the `test-pod.yaml` file so that it includes appropriate resource limits and requests; pod creation should also be allowed.

- **Enforcing namespace labeling with OPA Gatekeeper**:

  Enforcing namespace labeling is important for maintaining a consistent and organized Kubernetes environment. Labels provide a way to categorize and filter namespaces based on specific criteria, such as application type, environment, or ownership. OPA Gatekeeper can be used to enforce mandatory labeling of namespaces:

  I.   **Create a constraint template**: Create a file named `namespace-label-template.yaml` with the following content:

```
apiVersion: templates.gatekeeper.sh/v1beta1
kind: ConstraintTemplate
metadata:
  name: k8srequiredlabels
```

```
spec:
  crd:
    spec:
      names:
        kind: K8sRequiredLabels
        listKind: K8sRequiredLabelsList
        plural: k8srequiredlabels
        singular: k8srequiredlabels
  targets:
    - target: admission.k8s.gatekeeper.sh
      rego: |
        package k8srequiredlabels

        violation[{"msg": msg}] {
          input.review.object.kind == "Namespace"
          not input.review.object.metadata.labels[input.
parameters.label]
          msg := sprintf("Namespace must have the label '%v'",
[input.parameters.label])
        }
```

II.   Apply the constraint template to your cluster by running the following command:

```
$ kubectl apply -f namespace-label-template.yaml
```

III.   **Create a constraint:** Now, create a file named `namespace-label-constraint.yaml` with the following content:

```
apiVersion: constraints.gatekeeper.sh/v1beta1
kind: K8sRequiredLabels
metadata:
  name: namespace-labels
spec:
  parameters:
    label: "environment"
  match:
    kinds:
      - apiGroups: [""]
        kinds: ["Namespace"]
```

IV.   This constraint enforces that all namespaces must have a label named `environment`.

V. Apply the constraint to your cluster by running the following command:

```
$ kubectl apply -f namespace-label-constraint.yaml
```

VI. To test the constraint, create a file named `test-namespace.yaml` with the following content:

```
apiVersion: v1
kind: Namespace
metadata:
  name: test-namespace
```

VII. Try creating the test namespace by running the following command:

```
$ kubectl apply -f test-namespace.yaml
```

VIII. You should receive an error message indicating that the namespace must have the `environment` label. Update the `test-namespace.yaml` file so that it includes the required label; namespace creation should also be allowed.

- **Enforcing an image registry policy with OPA Gatekeeper**:

    Enforcing an image registry policy ensures that only images from trusted registries are allowed to be deployed in your Kubernetes cluster. This prevents potential security risks posed by deploying unknown or malicious container images. OPA Gatekeeper makes it easy to enforce such a policy across your entire cluster:

    I. **Create a constraint template**: Create a file named `allowed-repos-template.yaml` with the following content:

```
apiVersion: templates.gatekeeper.sh/v1beta1
kind: ConstraintTemplate
metadata:
  name: k8sallowedrepos
spec:
  crd:
    spec:
      names:
        kind: K8sAllowedRepos
        listKind: K8sAllowedReposList
        plural: k8sallowedrepos
        singular: k8sallowedrepos
  targets:
    - target: admission.k8s.gatekeeper.sh
      rego: |
        package k8sallowedrepos

        violation[{"msg": msg}] {
```

```
         input.review.object.kind == "Pod"
   container := input.review.object.spec.containers[_]
not startswith(container.image, input.parameters.repo)
msg := sprintf("Container image '%v' is not allowed. Only images
from the '%v' registry are permitted.", [container.image, input.
parameters.repo])
}
```

II.   Apply the constraint template to your cluster by running the following command:

```
$ kubectl apply -f allowed-repos-template.yaml
```

III.   **Create a constraint**: Now, create a file named namespace-label-constraint.
       yaml with the following content:

```
apiVersion: constraints.gatekeeper.sh/v1beta1
kind: K8sAllowedRepos
metadata:
  name: allowed-repos
spec:
  parameters:
    repo: "docker.io/your-organization"
  match:
    kinds:
      - apiGroups: [""]
        kinds: ["Pod"]
```

IV.   This constraint enforces that all container images must come from the specified registry.

V.   Apply the constraint to your cluster by running the following command:

```
$ kubectl apply -f namespace-label-constraint.yaml
```

VI.   To test the constraint, create a file named test-pod.yaml with the following content:

```
apiVersion: v1
kind: Pod
metadata:
  name: test-pod
spec:
  containers:
  - name: test-container
    Image. unauthorized-repo/nginx:latest
```

VII. Try to apply the test pod by running the following command:

```
$ kubectl apply -f test-pod.yaml
```

VIII. You should receive an error message indicating that the container image is not allowed. Update the `test-pod.yaml` file so that it uses an image from the allowed repository; pod creation should also be allowed.

- **Enforcing resource limits with OPA Gatekeeper**:

  Resource limits are crucial in maintaining the stability and performance of your Kubernetes cluster. They ensure that no single pod or container can consume an excessive amount of system resources, causing other applications to suffer. OPA Gatekeeper allows you to enforce resource limits across your entire cluster, ensuring that all deployed applications adhere to the specified constraints:

  I. **Create a constraint template**: Create a file named `resource-limits-template.yaml` with the following content:

```
apiVersion: templates.gatekeeper.sh/v1beta1
kind: ConstraintTemplate
metadata:
  name: k8sresourcelimits
spec:
  crd:
    spec:
      names:
        kind: K8sResourceLimits
        listKind: K8sResourceLimitsList
        plural: k8sresourcelimits
        singular: k8sresourcelimits
  targets:
    - target: admission.k8s.gatekeeper.sh
      rego: |
        package k8sresourcelimits

        violation[{"msg": msg}] {
          input.review.object.kind == "Pod"
          container := input.review.object.spec.containers[_]
          not container.resources.limits["cpu"]
          not container.resources.limits["memory"]
          msg := "Container must have CPU and memory limits specified."
        }
```

II.   Apply the constraint template to your cluster by running the following command:

```
$ kubectl apply -f resource-limits-template.yaml
```

III.   **Create a constraint**: Now, create a file named `resource-limits-constraint.yaml` with the following content:

```
apiVersion: constraints.gatekeeper.sh/v1beta1
kind: K8sResourceLimits
metadata:
  name: resource-limits
spec:
  match:
    kinds:
      - apiGroups: [""]
        kinds: ["Pod"]
```

IV.   This constraint enforces that all containers in a pod must have CPU and memory limits specified.

V.   Apply the constraint to your cluster by running the following command:

```
$ kubectl apply -f resource-limits-constraint.yaml
```

VI.   To test the constraint, create a file named `test-pod.yaml` with the following content:

```
apiVersion: v1
kind: Pod
metadata:
  name: test-pod
spec:
  containers:
  - name: test-container
    image. nginx
```

VII.   Try to apply the test pod by running the following command:

```
$ kubectl apply -f test-pod.yaml
```

VIII. You should receive an error message indicating that the container must have CPU and memory limits specified. Update the `test-pod.yaml` file so that it includes the required resource limits; pod creation should also be allowed.

These tutorials demonstrate how OPA Gatekeeper can be used to implement various security checks in a Kubernetes cluster, enforce object access control, and ensure compliance with best practices. By using OPA Gatekeeper, you can maintain a consistent security posture across your entire cluster and prevent potential issues before they arise.

Defense in depth

The dynamic nature of cloud-native applications, the ephemeral nature of containers, and the ever-increasing attack surface make securing these environments a critical priority. In this context, a defense-in-depth approach to cloud-native infrastructure security is essential.

Defense in depth is a comprehensive security strategy that relies on multiple layers of protection, ensuring that even if one layer is compromised, other layers remain intact. This layered approach is well suited for cloud-native environments, where the infrastructure is distributed across multiple components, such as VMs, containers, and serverless functions. By implementing security controls at each layer, organizations can better protect their applications and data from a wide range of threats and vulnerabilities.

The importance of securing the infrastructure in cloud-native environments cannot be overstated. As the foundation upon which applications and services are built, the infrastructure is a prime target for attackers looking to exploit vulnerabilities and gain unauthorized access. A security breach at the infrastructure level can lead to the compromise of multiple applications and services, potentially causing widespread damage and loss of data.

Furthermore, cloud-native environments often rely on APIs and microservices to facilitate communication between components. These APIs and microservices may expose sensitive data or functionality, increasing the attack surface and creating additional points of vulnerability. In this context, it is crucial to ensure that the infrastructure is properly secured, with robust access controls, network segmentation, and encryption in place.

The role of defense in depth in infrastructure security is to provide multiple layers of protection, making it more difficult for an attacker to breach the environment. In a cloud-native environment, this can be achieved by implementing security controls at various levels, such as the following:

- **Network security**: Network segmentation and firewall rules should be employed to isolate different components of the infrastructure, limiting the potential for lateral movement in the event of a breach. Additionally, traffic should be encrypted both in transit and at rest, ensuring that data is protected even if intercepted.

- **Identity and access management (IAM)**: Implementing a robust IAM strategy is crucial in a cloud-native environment. This involves defining roles and permissions, ensuring that users and services have the least privilege necessary to perform their tasks. Access to sensitive resources should be tightly controlled and monitored, with **multi-factor authentication** (**MFA**) and SSO employed where appropriate to enhance security.

- **Monitoring and logging**: Centralized monitoring and logging solutions should be implemented to collect data from all infrastructure components. This data can be analyzed for anomalies or potential threats, with alerts set up to notify administrators of any unusual activity. Regular log analysis can also help identify potential weaknesses in the infrastructure, allowing for proactive remediation.

- **Vulnerability management and patching**: Infrastructure components should be regularly scanned for vulnerabilities, with automated patch management processes in place to ensure timely updates. By staying up to date with the latest security patches, organizations can minimize the risk of known vulnerabilities being exploited.

- **Security incident response and remediation**: In the event of a security breach, a well-defined incident response plan is essential. This plan should outline the steps to be taken to contain the breach, investigate the cause, and remediate any damage. Regular security training for infrastructure teams can also help prevent incidents and ensure a swift, effective response if an incident occurs.

By implementing a defense-in-depth strategy for cloud-native infrastructure security, organizations can better protect their applications and data from a wide range of threats. This approach requires a combination of technology, processes, and people, working together to create a robust security posture. By investing in the necessary tools, resources, and training, organizations can significantly reduce the risk of security breaches and maintain the trust of their users and customers.

In conclusion, the importance of securing the infrastructure in cloud-native environments cannot be overstated. Defense in depth plays a crucial role in infrastructure security by providing multiple layers of protection, making it more challenging for attackers to breach the environment. By adopting a defense-in-depth approach, organizations can better protect their applications and data from a wide range of threats and vulnerabilities, ensuring the ongoing resilience and security of their cloud-native infrastructure.

Infrastructure components in cloud-native environments

In cloud-native environments, various infrastructure components work together to provide the foundation for applications and services. These components include VMs, containers, serverless computing, networking components such as VPCs, subnets, load balancers, and ingress controllers, and storage services such as block storage, object storage, and databases. To effectively implement defense in depth, it is crucial to apply security best practices and the principle of least privilege across each of these components. Let's begin by understanding the components and their architecture.

Compute components – virtual machines, containers, and serverless computing

VMs, containers, and serverless computing are all popular compute options in cloud-native environments. Each of these options has unique security considerations, and defense in depth should be applied accordingly:

- For VMs, ensure that the host operating system and any guest operating systems are properly patched and secured. Implement strict access controls to limit user access to VMs and apply the principle of least privilege to restrict permissions. VM isolation should be enforced, and VM traffic should be monitored for suspicious activity.

- For containers, use secure base images and scan images for vulnerabilities. Apply container isolation technologies such as gVisor or Kata Containers to limit the blast radius of potential attacks. Follow the principle of least privilege by limiting container permissions and use namespace isolation to segregate workloads.

- For serverless computing, implement strict function access controls and apply the principle of least privilege. Monitor function invocations and logs for anomalies and limit the execution time of functions to reduce the potential for abuse.

One example of applying defense in depth in VMs is ensuring secure configuration and patch management. An organization can implement automated patch management systems to routinely update both the host and guest operating systems, reducing the attack surface. Additionally, hardening the VMs by disabling unnecessary services and following best practices such as the CIS Benchmarks can provide an extra layer of security.

As an example, a financial services company experienced a security breach because an unpatched vulnerability in the guest OS allowed an attacker to gain unauthorized access. By implementing an automated patch management system and hardening the VMs, the company was able to minimize the risk of future breaches.

In a containerized microservices architecture, defense in depth can be applied by combining multiple security layers, such as network segmentation, container isolation, and strict RBAC policies.

For example, an organization running a containerized microservices application might use Kubernetes network policies to isolate individual microservices and limit communication between them. Additionally, the organization could employ gVisor or Kata Containers for enhanced container isolation and use strict RBAC policies to control access to Kubernetes resources.

Anecdotally speaking, a tech company experienced a security incident in which an attacker gained access to a single container and attempted to move laterally within the cluster. By implementing defense in depth with network segmentation, container isolation, and RBAC, the company was able to limit the attacker's access and prevent further damage.

Networking components – VPCs, subnets, load balancers, and ingress controllers

Applying defense in depth to networking components involves multiple layers of security, including network segmentation, traffic encryption, and access control:

- For VPCs and subnets, create a segmented network architecture that isolates different application tiers and limits lateral movement in case of a breach. Implement strict network **access control lists (ACLs)** and security groups to restrict traffic between components.

- For load balancers and ingress controllers, configure secure SSL/TLS termination and enforce strict access controls. Apply the principle of least privilege to limit access to these components and monitor traffic for anomalies.

A practical example of defense in depth in networking is applying network segmentation in a multi-tier application architecture. By isolating each tier in its own subnet and applying strict network ACLs and security groups, an organization can limit lateral movement if one tier is compromised.

In an e-commerce application, for example, separate subnets can be created for the frontend, application, and database tiers. This segmentation ensures that even if the frontend is compromised, the attacker would have difficulty reaching the application or database tiers without bypassing additional security controls.

Storage services – block storage, object storage, and databases

Protecting storage services in a cloud-native environment is crucial for safeguarding sensitive data:

- For block storage and object storage, implement encryption at rest and in transit. Use strict access controls and IAM policies to limit access to these storage resources, following the principle of least privilege.

- For databases, configure secure database connections and implement strong authentication mechanisms, such as password policies and MFA. Limit database access to authorized users and services and monitor database activity for signs of unauthorized access or data exfiltration.

A cloud-native organization might store sensitive customer data in object storage, such as AWS S3 or Google Cloud Storage. By implementing fine-grained access controls using IAM policies and ACLs, the organization can ensure that only authorized users and services can access the data. The principle of least privilege can be applied by granting users the minimum permissions necessary to perform their tasks.

In a healthcare application, for example, patient records are stored in object storage; an IAM policy can be configured to allow specific users and services to access only the records they need. This prevents unauthorized access and limits the potential damage in case of a breach.

By applying defense in depth and the principle of least privilege across each of these infrastructure components, organizations can build a more secure foundation for their cloud-native applications and services. This approach ensures that even if one layer of security is compromised, other layers remain in place to protect the environment.

Now that we have got a fair idea of the cloud-native workloads and components that are widely used and are frequently observed in the wild, let's try to understand how we can monitor these environments in real time to find security vulnerabilities and threats in the production cluster.

Falco – real-time monitoring for cloud workloads

Falco is an open source, cloud-native runtime security project that focuses on detecting anomalous behavior in your applications and infrastructure. Created by Sysdig and now a part of the **Cloud Native Computing Foundation** (**CNCF**), Falco is built on top of the Linux kernel and leverages the **Extended Berkeley Packet Filter** (**eBPF**) to provide real-time security monitoring.

Falco's main purpose is to alert users to suspicious activity in their cloud-native environments, including Kubernetes clusters, containers, and applications. By monitoring system calls and events at the kernel level, Falco can detect and notify security teams about potential security breaches or policy violations in real time.

Some of the most widely used applications of Falco are as follows:

- **Intrusion and breach detection**: Falco can detect attempts to access sensitive files, unauthorized process execution, or attempts to exploit known vulnerabilities. By detecting these intrusions, Falco enables organizations to respond quickly to potential threats and minimize damage.

- **Compliance and audit**: Falco can help organizations maintain compliance with security standards and regulations by continuously monitoring their environment for violations of security policies. Falco can generate detailed audit logs, providing valuable insights for incident response and forensic analysis.

- **Kubernetes security**: Falco can monitor Kubernetes API server events, detecting unauthorized changes to the cluster's configuration or attempts to create malicious resources. This enables organizations to maintain control over their Kubernetes clusters and protect against unauthorized access or manipulation.

- **Container security**: Falco can detect unauthorized container access, privilege escalation, or unexpected network connections. This helps organizations maintain a strong security posture for their containerized applications and minimize the risk of container-based attacks.

- **Application security**: By monitoring application-level events, Falco can detect potential security issues such as SQL injection, **cross-site scripting** (**XSS**), or **remote code execution** (**RCE**) attempts.

Let's learn how Falco enables detecting a real-time attack or even a pentesting engagement.

Security attack chains monitored by Falco

Let's understand this by breaking it down into individual steps of a pentesting exercise:

- **Lateral movement**: Falco can detect attempts to move laterally within a Kubernetes cluster or container environment, alerting security teams to potential insider threats or targeted attacks

- **Privilege escalation**: Falco can identify attempts to escalate privileges, such as exploiting a vulnerable application or container to gain root access to the underlying system

- **Data exfiltration**: By monitoring network connections, Falco can detect attempts to exfiltrate sensitive data from your environment, such as an attacker attempting to transfer data to a remote server

- **Command-and-control (C2) communication**: Falco can identify suspicious network connections that may indicate communication with a command-and-control server, which is often a sign of an ongoing attack

Now that we understand the benefits and rationale of employing Falco in a production environment, let's learn how to implement it.

Getting started with Falco

To get started with Falco, you can install it on a host, within a container, or as a Kubernetes DaemonSet. Falco comes with a set of predefined rules that cover common security concerns, but you can also create custom rules tailored to your specific environment and security requirements. These rules are written in YAML and can be updated dynamically without restarting Falco.

Falco integrates with various security and incident response tools, such as Prometheus, Elasticsearch, and Grafana, making it easier to incorporate Falco's monitoring capabilities into your existing security stack. Falco provides many ways to integrate with your existing cloud-native workflow. Let's explore a few of those.

Integrating Falco into your cloud-native workflow and environment

Integrating Falco into your cloud-native environment is a straightforward process, and it can be deployed in various ways, such as on a host, within a container, or as a Kubernetes DaemonSet:

- **Host installation**: You can install Falco directly on your host by using pre-built packages or by building it from source. Pre-built packages are available for popular Linux distributions such as Debian, Ubuntu, CentOS, and RHEL.

- **Container deployment**: Falco can be deployed as a Docker container or using container orchestration platforms such as Kubernetes. This allows you to monitor container runtime activity and enforce security policies across your containerized applications.

- **Kubernetes DaemonSet**: Deploying Falco as a Kubernetes DaemonSet ensures that a Falco instance runs on each node in your cluster. This provides comprehensive coverage of your entire Kubernetes environment and allows for monitoring both the control plane and worker nodes.

Now that Falco has been integrated into the workflow, let's learn how to use the tool for real-time threat detection.

Configuring Falco rules for monitoring and alerting

Falco comes with a set of predefined rules that cover common security concerns. These rules are written in YAML and are designed to be human-readable and easy to modify. To configure Falco rules, follow these steps:

1. Locate the `falco_rules.yaml` file, which contains the default ruleset. You can find this file in the `/etc/falco` directory for host installations, or within the Falco container for containerized deployments.

2. Create a new custom rules file, such as `custom_rules.yaml`, to store your custom rules. This file should be placed in the same directory as the `falco_rules.yaml` file.

3. Edit the custom rules file using a text editor, adding or modifying rules as needed. Each rule should include the following components:

   - `rule`: A unique name for the rule

   - `condition`: A Boolean expression that defines the conditions under which the rule should trigger

   - `output`: A message to be displayed when the rule triggers

   - `priority`: The severity level of the rule, such as "emergency," "alert," "critical," "error," "warning," "notice," "info," or "debug".

4. Update the `falco.yaml` configuration file so that it includes your custom rules file. Add the `/etc/falco/custom_rules.yaml` line under the `rules_file` section.

5. Restart the Falco service to apply the new rules.

Now that we have created a sample Falco rule, let's learn about its best practices so that you can use this tool in the production environment of your cloud-native workload.

Best practices for Falco deployment and configuration

Here are the best practices:

- **Use a centralized logging solution**: Integrate Falco with a centralized logging system, such as Elasticsearch or Splunk, to consolidate and analyze the logs generated by Falco.

- **Monitor Falco logs for anomalies**: Regularly review Falco logs to identify any unexpected or suspicious activity. This can be done manually or by using log analysis tools and alerting systems.

- **Implement least privilege**: Configure Falco to run with the least privilege necessary to perform its monitoring functions. Avoid running Falco as the root user whenever possible.

- **Regularly update Falco**: Ensure that Falco is always up to date with the latest security patches and rule updates. Subscribe to the Falco mailing list or GitHub repository to stay informed about new releases and updates.

- **Use network segmentation**: Isolate Falco instances and the systems they monitor by using network segmentation, reducing the attack surface, and limiting the potential impact of a security breach.

Often, organizations choose to use commercial monitoring products for threat detection or threat hunting, or might already have an existing tool in place, since there cannot be too many security tools in place. You can also use Falco as a companion threat detection tool. Let's learn how to integrate Falco with existing monitoring tools.

Using Falco with existing monitoring and logging tools

Falco can be easily integrated with existing monitoring and logging tools, such as Prometheus, Elasticsearch, Grafana, and Splunk. This allows you to correlate Falco alerts with other system metrics and logs, providing a more comprehensive view of your environment's security posture.

To integrate Falco with these tools, you can use output plugins, such as these:

- **Falco-Prometheus**: This plugin exports Falco metrics in a format that can be scraped by Prometheus, a popular monitoring and alerting tool.

- **Falco-Elasticsearch**: This plugin enables Falco to send events directly to Elasticsearch, a powerful search and analytics engine. By ingesting Falco events into Elasticsearch, you can visualize and analyze Falco data alongside other log data in Kibana, a web-based user interface for Elasticsearch.

- **Falco-Grafana**: This plugin allows you to create Grafana dashboards that visualize Falco metrics and alerts. By combining Falco data with other system and application metrics in Grafana, you can create a comprehensive monitoring and alerting solution for your cloud-native environment.

- **Falco-Splunk**: This plugin forwards Falco events to Splunk, a popular log management and analytics platform. By integrating Falco with Splunk, you can create alerts, dashboards, and reports based on Falco data and correlate it with other log sources.

Using Falco in large-scale environments

In large-scale environments, it's crucial to ensure that Falco is deployed and configured optimally for performance, reliability, and scalability. Here are some recommendations for using Falco in such environments:

- **Use Kubernetes DaemonSets**: Deploy Falco as a Kubernetes DaemonSet to ensure that a Falco instance is running on every node in the cluster. This provides comprehensive coverage and allows you to monitor all the containers running in the environment.

- **Use horizontal pod autoscaling (HPA)**: Configure HPA for your Falco instances to scale the number of instances based on CPU utilization or custom metrics, ensuring adequate resources are available for monitoring and alerting.

- **Implement resource limits and requests**: Set appropriate resource limits and requests for your Falco instances to ensure they have adequate CPU and memory resources, and to prevent them from consuming excessive resources in the cluster.

- **Centralize log collection and analysis**: Integrate Falco with a centralized log management and analysis platform, such as Elasticsearch or Splunk, or a managed service such as AWS CloudWatch or Google Stackdriver. This helps you manage and analyze large volumes of log data generated by Falco instances.

- **Regularly review and update Falco rules**: Continuously review and update Falco rules to ensure they are relevant and effective in detecting security threats in your environment. Additionally, consider implementing custom rules tailored to your organization's specific security requirements.

By following these best practices, you can successfully deploy and use Falco to monitor and secure your cloud-native infrastructure in large-scale environments.

In conclusion, Falco is a powerful and flexible tool for real-time security monitoring in cloud-native environments. By detecting potential security issues and policy violations early, Falco helps organizations maintain a strong security posture and respond quickly to threats. Its wide range of use cases and ability to monitor various attack chains make it an essential component of a comprehensive cloud-native security strategy.

Summary

In this chapter, we explored various aspects of securing the infrastructure in cloud-native environments. By adhering to best practices and leveraging the right tools and platforms, you can create a strong foundation for protecting your applications and data in the cloud.

We started by discussing the approach to object access control, emphasizing the importance of Kubernetes network policies and how they can be used to secure communication between pods within a cluster. We also covered the role of Calico, a powerful networking and network security solution that can enhance the native capabilities of Kubernetes network policies.

Next, we delved into the principles of authentication and authorization, highlighting the role of Kubernetes' native features in managing access control. We also discussed the importance of using tools such as OPA Gatekeeper to enforce policies within the cluster and ensure that only authorized actions are permitted.

Lastly, we examined the concept of defense in depth for cloud-native infrastructure. We emphasized the significance of securing various infrastructure components, such as virtual machines, containers, networking components, and storage services. We also provided a detailed explanation of Falco, a powerful tool for real-time security monitoring in cloud-native environments. We discussed its use cases, best practices for deployment and configuration, and how it can be integrated with existing monitoring and logging tools.

By understanding and implementing the various security measures and tools discussed in this chapter, you can significantly enhance the security posture of your cloud-native environment and protect your applications and data from potential threats and vulnerabilities.

In the next chapter, we will be learning on how to perform security operations and incident response now that we have learned to secure the infrastructure and put security controls in place.

Quiz

Answer the following questions to test your knowledge of this chapter:

- How do Kubernetes network policies and Calico work together to provide a comprehensive approach to object access control, and what are the key benefits of using Calico over native Kubernetes network policies?

- In the context of authentication and authorization in cloud-native environments, what are the advantages of using tools such as OPA Gatekeeper in addition to Kubernetes' native features for access control, and how can these tools be best utilized?

- How can defense-in-depth strategies be applied to various infrastructure components in cloud-native environments, such as virtual machines, containers, networking components, and storage services, and what role does the principle of least privilege play in these strategies?

- In the context of real-time security monitoring, what are the key use cases for Falco, and how can organizations effectively integrate it into their cloud-native workflows for large-scale environments?

- Considering the various security aspects discussed in this chapter, how can organizations strike a balance between maintaining a high level of security and ensuring flexibility and ease of use for developers and operators in cloud-native environments?

Further readings

To learn more about the topics that were covered in this chapter, take a look at the following resources:

- `https://kubernetes.io/docs/concepts/services-networking/network-policies/`

- `https://github.com/ahmetb/kubernetes-network-policy-recipes`

- `https://docs.projectcalico.org/getting-started/kubernetes/`

- `https://open-policy-agent.github.io/gatekeeper/website/docs/`

- `https://kubernetes.io/docs/reference/access-authn-authz/authorization/`

- `https://kubernetes.io/docs/reference/access-authn-authz/authentication/`

- `https://nvlpubs.nist.gov/nistpubs/SpecialPublications/NIST.SP.800-53r5.pdf`

- `https://containerjournal.com/topics/container-ecosystems/defence-in-depth-for-cloud-native-applications/`

- `https://github.com/falcosecurity/falco`

7

Cloud Security Operations

Cloud security operations are crucial in safeguarding the confidentiality, integrity, and availability of cloud-native applications. As reliance on cloud-native infrastructures grows, it is imperative to implement effective security measures and consistently monitor applications to detect and mitigate potential threats. This chapter offers practical insights and tools for establishing and maintaining a robust cloud security operations process.

We'll explore innovative techniques for collecting and analyzing data points, including centralized logging, cloud-native observability tools, and monitoring with Prometheus and Grafana. You'll learn how to proactively detect potential threats and create alerting rules and webhooks within various platforms to automate incident response.

Next, we'll examine automated security lapse findings using **Security Orchestration, Automation, and Response (SOAR)** platforms, integrating security tools, and building a security automation playbook. This will help streamline security operations and minimize human intervention.

We'll guide you through defining runbooks to handle security incidents and develop a security incident response plan. By conducting regular drills and refining runbooks, your team will be prepared to tackle real-life scenarios.

Lastly, we'll discuss the investigation and reporting of security incidents, focusing on collaboration and using metrics and **Key Performance Indicators (KPIs)** to drive continuous improvement in security operations. By the end of this chapter, you'll be well equipped to establish and maintain a comprehensive cloud security operations process, ensuring the security and resilience of your cloud-native applications.

In this chapter, we're going to cover the following main topics:

- Novel techniques in sourcing data points
- Creating alerting and webhooks within different platforms
- Automated security lapse findings
- Defining runbooks to handle security incidents

Technical requirements

The technical requirements for this chapter are as follows:

- Elasticsearch v7.13.0

- Fluentd v1.15.1

- Kibana v8.7.0

- Prometheus v2.44.0

- Helm v3.12.0

- Kubernetes v1.27

Novel techniques in sourcing data points

In this section, we will explore innovative approaches to collecting anf analyzing data from cloud-native platforms such as Kubernetes and cloud environments. Gathering data points effectively is essential for detecting potential threats and ensuring the security of your cloud-native applications. The techniques and tools we will discuss in this section will help you gain deeper insights into your application's behavior and security posture.

We will begin by discussing centralized logging using the EFK stack within Kubernetes. This approach will enable you to aggregate and analyze logs from multiple sources, providing a comprehensive view of your application's activities.

Centralized logging with the EFK stack

Let's begin by discussing, in detail, the process of implementing centralized logging with the EFK stack in a Kubernetes environment. We'll cover the architecture, configuration, and deployment of the EFK stack and explain how to create visualizations and dashboards to monitor security events and detect potential threats.

Architecture overview

The ELK stack comprises three main components:

- **Elasticsearch**: A distributed, RESTful search and analytics engine designed for scalability, resilience, and real-time data analysis.

- **Fluentd**: A lightweight, flexible, and cloud-native log collector and processor that aggregates and forwards logs from various sources to Elasticsearch. It is designed for high performance and extensibility, with a rich ecosystem of plugins for data collection, filtering, and output.

- **Kibana**: A web-based user interface for visualizing and exploring Elasticsearch data, creating dashboards, and managing the Elastic Stack.

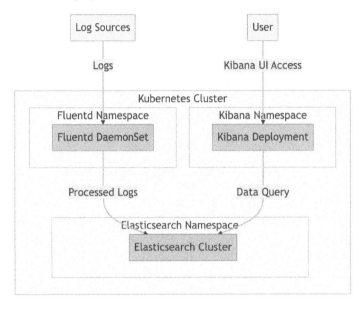

Figure 7.1 – EFK stack architecture in Kubernetes

Kubernetes generates a significant amount of log data from its cluster components, such as nodes, pods, and containers. Centralizing and analyzing these logs is crucial for detecting security threats, misconfigurations, and policy violations.

Setting up the EFK stack on Kubernetes

Deploying the **Elasticsearch, Fluentd, and Kibana** (**EFK**) stack on a Kubernetes cluster involves several steps. This tutorial will provide a step-by-step guide using Helm, a Kubernetes package manager, as it simplifies the deployment and management of applications on Kubernetes.

Before you start, make sure you have the following prerequisites installed:

- A running Kubernetes cluster
- Helm v3
- `kubectl`, configured to interact with your Kubernetes cluster
- An Ingress Controller deployed in your cluster environment

> **Important note**
>
> Always ensure that your Kubernetes cluster is secured and that access is controlled via **role-based access control (RBAC)**. You should limit who can interact with your cluster and what permissions they have.

To deploy Elasticsearch in Kubernetes

The first component we will deploy is Elasticsearch. Elasticsearch is a search and analytics engine that will store and index the logs shipped by Fluentd:

1. Add the elastic Helm chart repo:

    ```
    $ helm repo add elastic https://helm.elastic.co
    ```

2. Update your Helm chart repositories:

    ```
    $ helm repo update
    ```

3. Create a values chart for Elasticsearch:

    ```
    ---
    protocol: http
    httpPort: 9200
    transportPort: 9300

    service:
      labels: {}
      labelsHeadless: {}
      type: LoadBalancer
      nodePort: ""
      annotations: {}
      httpPortName: http
      transportPortName: transport
      loadBalancerIP: ""
      loadBalancerSourceRanges: []
      externalTrafficPolicy: ""
    ```

The reason we use the LoadBalancer service type is to expose the Elasticsearch service externally so that other services can access Elasticsearch.

4. Deploy Elasticsearch using a Helm chart:

```
$ helm install elasticsearch elastic/elasticsearch --version
7.13.0 -f evalues.yaml
```

5. Note the Elasticsearch external IP address, as we will need that later.

> **Important note**
>
> The preceding command deploys a basic configuration of Elasticsearch, which may not be suitable for production environments. For a production setup, you should consider factors such as resource requirements, storage class configurations, and high availability. Always ensure to use PersistentVolumes for Elasticsearch data storage to prevent data loss when pods are rescheduled.

With Elasticsearch up and running, let us look at provisioning Fluentd in our Kubernetes cluster.

To deploy Fluentd in Kubernetes

Fluentd will collect, transform, and ship the logs to Elasticsearch:

1. Fluentd runs as a DaemonSet, and you can download the manifest file from this GitHub URL:

   ```
   https://github.com/fluent/fluentd-kubernetes-daemonset/blob/
   master/fluentd-daemonset-elasticsearch-rbac.yaml
   ```

2. Change the FLUENT_ELASTICSEARCH_HOST value to the external IP address of the Elasticsearch service.

3. Set the value for FLUENTD_SYSTEMD_CONF as disable.

4. Run the following command:

   ```
   $ kubectl create -f fluentd-daemonset-elasticsearch-rbac.yaml
   ```

> **Important note**
>
> The Fluentd configuration needs to be tailored to your specific logging needs. The configuration defines input, filter, and output plugins. The input plugins define the log sources, filter plugins transform the logs, and output plugins define where to forward the logs. The chart by Bitnami by default forwards logs to an Elasticsearch instance running in the same Kubernetes cluster.

Finally, let us provision Kibana to orchestrate the EFK stack in our Kubernetes cluster.

To deploy Kibana in Kubernetes

Finally, we will deploy Kibana. Kibana provides a user-friendly web interface to search and visualize the logs stored in Elasticsearch:

1. Use the following manifest and store it in `kibana-values.yaml`:

```
---
elasticsearchHosts: "http://10.116.200.220:9200"
replicas: 1
image: "docker.elastic.co/kibana/kibana"
imageTag: "7.13.0"
imagePullPolicy: "IfNotPresent"
resources:
  requests:
    cpu: "1000m"
    memory: "1Gi"
  limits:
    cpu: "1000m"
    memory: "1Gi"
healthCheckPath: "/api/status"
httpPort: 5601
service:
  type: LoadBalancer #ClusterIP
  loadBalancerIP: ""
  port: 5601
  nodePort: ""
  labels: {}
  annotations: {}
    # cloud.google.com/load-balancer-type: "Internal"
    # service.beta.kubernetes.io/aws-load-balancer-internal:
0.0.0.0/0
    # service.beta.kubernetes.io/azure-load-balancer-internal:
"true"
    # service.beta.kubernetes.io/openstack-internal-load-
balancer: "true"
    # service.beta.kubernetes.io/cce-load-balancer-internal-vpc:
"true"
  loadBalancerSourceRanges: []
    # 0.0.0.0/0
  httpPortName: http
```

Change `elasticsearchHosts` to the IP address of `elasticsearch`.

2. Deploy Kibana using the Helm chart:

```
$ helm install kibana -version 7.13.0 elastic/kibana -f kibana-
values.yaml
```

> **Important note**
>
> By default, Kibana runs on port 5601 and isn't exposed outside the Kubernetes cluster. To access Kibana, you can set up port forwarding with kubectl port-forward or set up an Ingress resource. With the EFK stack deployed and configured, you're now ready to start exploring and analyzing the logs from your Kubernetes cluster. This will give you valuable insights into the operation and performance of your cluster, help you troubleshoot and resolve issues, and enhance the security and compliance of your environment.

Collecting and processing logs

Fluentd collects logs from various sources, such as Kubernetes nodes, pods, and containers, using input plugins. Some commonly used input plugins for Kubernetes include tail, forward, and syslog.

Once logs are collected, Fluentd processes and transforms them using filter plugins, which can parse, mutate, and enrich log data before forwarding it to Elasticsearch.

Some examples of filter plugins are as follows:

- **Parser**: Parses unstructured log data using regular expressions or built-in parsers such as JSON, CSV, and Apache

- **Record Transformer**: Modifies log data, such as adding, renaming, or removing fields

- **GeoIP**: Enriches log data with geographical information based on IP addresses

- **Detect Exceptions**: Aggregates multi-line exception stack traces into a single log record

With Fluentd, you can efficiently collect, process, and forward logs from your Kubernetes environment, making it easier to analyze and monitor your applications and infrastructure. Its extensible architecture and rich plugin ecosystem allow for easy customization and integration with various log sources and outputs:

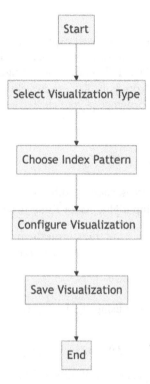

Figure 7.2 – Kibana visualization process flow

With our newfound understanding of how to create visualizations within Kibana, let us take the use case of our EFK stack a step further by learning about other security threat detections and configuring alerting and monitoring techniques for our Kubernetes cluster.

Configuring alerting and monitoring

After logs are processed and stored in Elasticsearch, you can use Kibana to create visualizations and dashboards that provide insights into your application's security posture. This includes detecting unauthorized access attempts, monitoring resource consumption, and identifying suspicious network traffic patterns:

1. **Create visualizations**: Use Kibana to create various types of visualizations, such as bar charts, pie charts, and heatmaps, to represent log data and security events

2. **Build dashboards**: Combine multiple visualizations into a single dashboard to monitor key security metrics and events. Customize the dashboard layout, filters, and time range to suit your specific requirements

3. **Set up alerts**: Configure alerts in Kibana to notify you when specific security events or threshold breaches occur. Alerts can be sent via email, Slack, or other communication channels, allowing your team to respond quickly to potential threats

Securing the EFK stack

It's crucial to secure the EFK stack components to prevent unauthorized access and data breaches. Here are some best practices for securing your EFK stack deployment:

- **Enable security features**: Use Elasticsearch's built-in security features, such as RBAC, node-to-node encryption, and TLS/SSL encryption for client connections

- **Secure Kibana**: Configure Kibana to use TLS/SSL encryption and integrate it with Elasticsearch's RBAC to restrict user access to specific data and features

- **Network segmentation**: Deploy the EKF stack components within a dedicated Kubernetes namespace and use network policies to restrict traffic between namespaces and applications

- **Logstash security**: Configure Fluentd input plugins to use secure protocols, such as HTTPS or SSL/TLS-encrypted syslog, to prevent data tampering during log ingestion

Now that we have provisioned, secured, and deployed a scalable EFK stack in our Kubernetes environment, let us look into some of the security findings in our cloud-native environment, by virtue of the deployed EFK stack.

Detecting security threats, misconfigurations, and policy violations

Detecting security threats, misconfigurations, and policy violations in your cloud-native environment is crucial for maintaining a secure and compliant infrastructure. By proactively monitoring and analyzing logs using the EFK stack, you can identify potential vulnerabilities and attacks, and take appropriate actions to prevent or mitigate any damage. Some of the main reasons to implement this are as follows:

- **Maintain regulatory compliance**: Monitoring for policy violations ensures that your environment remains compliant with organizational and regulatory requirements. Non-compliance can result in fines, penalties, and reputational damage.

- **Improve the overall security posture**: By detecting misconfigurations and security threats, you can continuously improve your environment's security posture, reducing the attack surface and making it more difficult for attackers to exploit vulnerabilities.

- **Efficient incident response**: Quick detection of security incidents enables your team to respond faster and more effectively, minimizing the potential impact on your infrastructure and business operations.

- **Optimize resource utilization**: Monitoring for excessive resource consumption allows you to identify and resolve performance bottlenecks, ensuring optimal resource utilization in your environment.

By implementing the EFK stack for centralized logging in your infrastructure, you can proactively monitor and detect potential security incidents. Some examples of security threats, misconfigurations, and policy violations that can be detected through log analysis are as follows:

- **Unauthorized access attempts**: Monitor logs for failed authentication attempts and potential brute-force attacks on the Kubernetes API server or other components

- **Privilege escalation**: Detect attempts to exploit misconfigured RBAC settings or escalate privileges within the cluster

- **Excessive resource consumption**: Identify abnormal resource usage by specific pods, which may indicate potential **Denial of Service (DoS)** attacks or resource exhaustion

- **Suspicious network traffic patterns**: Analyze logs to discover unusual connections between cluster nodes or external IP addresses, which may suggest network intrusions or data exfiltration attempts

- **Security misconfigurations**: Detect instances of improper security configurations, such as exposed secrets, insecure communication protocols, or weak encryption settings

- **Policy violations**: Identify the use of non-compliant container images, deployments that violate network segmentation rules, or other violations of organizational or regulatory policies

Let us look into some of the attack vectors we discussed previously that could be detected using our EFK stack.

Detecting unauthorized access attempts

Detecting unauthorized access attempts is crucial to maintaining the security of your Kubernetes environment. Early detection of such attempts can help prevent data breaches, protect sensitive information, and maintain compliance with regulatory requirements. Security engineers should care about detecting unauthorized access attempts for the following reasons:

- Unauthorized access can lead to data breaches, exposing sensitive information and causing reputational damage, financial loss, and legal consequences

- Detecting unauthorized access attempts helps security engineers identify weak security configurations, such as insufficient authentication or access control mechanisms, and improve the overall security posture of the infrastructure

- Timely detection and response to unauthorized access attempts can prevent attackers from exploiting vulnerabilities and gaining a foothold in the environment, reducing the likelihood of a successful attack

The EFK stack can be used to detect unauthorized access attempts by monitoring, processing, and analyzing logs from various sources within your Kubernetes environment. Here's a high-level overview of how the EFK stack components work together to detect unauthorized access attempts:

1. Fluentd collects logs from Kubernetes nodes, pods, and containers, including logs related to authentication and access control events.

2. Fluentd processes and enriches the logs, parsing them into structured JSON format and adding relevant metadata.

3. Fluentd forwards the processed logs to Elasticsearch for indexing and storage.

4. Kibana provides a web-based user interface for visualizing and analyzing logs stored in Elasticsearch, allowing you to create custom dashboards and alerts for monitoring unauthorized access attempts.

Let's take a hands-on look at how we can detect the aforementioned security threats in your Kubernetes environment. These tutorials assume you have a working Kubernetes environment with the EFK stack already deployed. If you haven't deployed the EFK stack, please refer to the previous sections on setting up Elasticsearch, Fluentd, and Kibana.

Implementing the EFK stack to detect unauthorized access attempts

Please note, these tutorials are only meant to be guides. Given that each cluster environment is different, please ensure that you run the code/use the right FQDN for your service name or pod deployment:

1. **Configure Fluentd to collect relevant logs**: Ensure Fluentd is configured to collect logs from Kubernetes API servers, nodes, and other components responsible for authentication and access control. In your Fluentd configuration, use the `kubernetes_metadata` filter plugin to add Kubernetes-specific metadata to the logs:

```
<source>
  @type tail
  path /var/log/containers/*.log
  pos_file /var/log/fluentd-containers.log.pos
  tag kubernetes.*
  read_from_head true
  <parse>
    @type json
    time_format %Y-%m-%dT%H:%M:%S.%NZ
  </parse>
</source>
```

```
<filter kubernetes.**>
  @type kubernetes_metadata
</filter>
```

2. **Create Kibana visualizations and dashboards**: Log in to Kibana and create visualizations for monitoring unauthorized access attempts. Open your web browser and navigate to your Kibana instance's URL. The URL will usually be in the format `http://<kibana-service-ip>:<kibana-service-port>`. Once you've accessed Kibana, you'll see the Kibana home page:

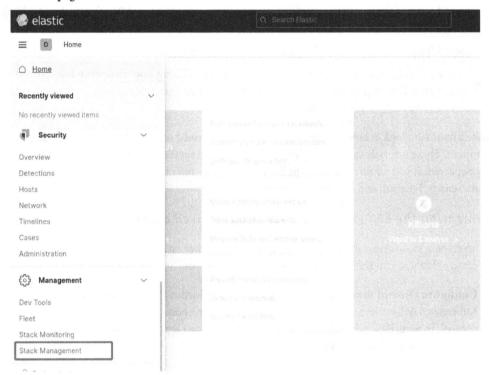

Figure 7.3 – Kibana dashboard

3. **Create an index pattern**: To create visualizations and dashboards, you first need to define an index pattern in Kibana that matches the Elasticsearch indices where your logs are stored. From the Kibana home page, click on the **Stack Management** tab in the left-hand menu, and then click **Index Patterns**. Click **Create index pattern** and enter an index pattern that matches your log indices (e.g., `fluentd-*`). Click **Next step** and select the time field for your logs, such as `@timestamp`. Finally, click **Create index pattern**:

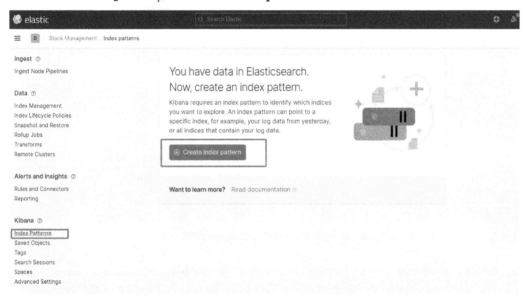

Figure 7.4 – Kibana index pattern

4. **Navigate to Visualize**: To create a new visualization, click on the **Visualize** tab in the left-hand menu. Click **Create visualization** to start creating a new visualization:

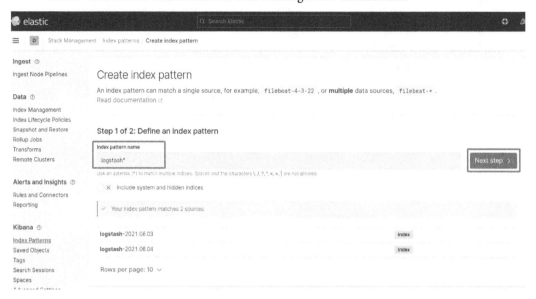

Figure 7.5 – Visualizing index patterns in Kibana

5. **Select a visualization type**: Choose a visualization type that best represents the data you want to display. To monitor failed authentication attempts, a bar chart or line chart might be suitable. Click on the desired visualization type to proceed.

6. **Configure the visualization**: Now you'll configure the visualization by defining the X-axis (horizontal) and Y-axis (vertical) data. For our example of monitoring failed authentication attempts, follow these steps:

    I.   Under the **Metrics** section, click on **Y-axis** and select **Count** as the aggregation type. This will display the number of failed authentication attempts on the Y-axis.

    II.   Under the **Buckets** section, click on **X-axis** and choose **Date Histogram** as the aggregation type. This will display the time on the X-axis. Choose the **@timestamp** field and an appropriate time interval (e.g., 1m for one-minute intervals).

III. To further group the data by the source IP address or username, click **Add sub-buckets** under the **Buckets** section. Select **Split series** and choose **Terms** as the aggregation type. Then, select the field you want to group by, such as `source_ip` or `user_name`. Set the **Order by** option to **Descending** and **Size** to a suitable value (e.g., 5 to display the top five IP addresses or usernames):

Figure 7.6 – Configuring custom visualizations

7. **Save the visualization**: Once you've configured the visualization, click the **Save** button at the top of the page. Give your visualization a descriptive name (e.g., **Failed Authentication Attempts**) and click **Save**.

8. **Create a dashboard**: Now that you've created a visualization, you can add it to a dashboard. Click on the **Dashboard** tab in the left-hand menu, and then click **Create dashboard**. Click **Add** at the top of the page and select your **Failed Authentication Attempts** visualization to add it to the dashboard.

9. **Save the dashboard**: Give your dashboard a descriptive name (e.g., **Security Monitoring**) and click **Save**. You can now use this dashboard to monitor unauthorized access attempts in real time.

10. **Customize the dashboard**: Dashboards in Kibana are highly customizable, allowing you to add multiple visualizations, adjust the layout, and apply filters to focus on specific data. To add more visualizations, click **Add** and select the visualizations you want to include. You can also resize and rearrange visualizations by dragging and dropping them on the dashboard.

11. **Apply filters**: Filters help you narrow down the data displayed in the dashboard. To apply a filter, click the **Add filter** button at the top of the dashboard. You can create filters based on specific fields (e.g., source IP address, username, or Kubernetes namespace) and conditions (e.g., equals, does not equal, is one of, or is not one of). For example, you can create a filter to display only failed authentication attempts from a specific IP address or within a specific Kubernetes namespace:

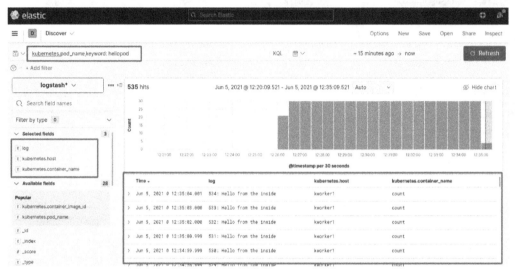

Figure 7.7 – Applying filters for visualization

12. **Set auto-refresh and time range**: To keep your dashboard up to date, you can configure it to refresh automatically at regular intervals. Click the **Refresh** button at the top of the dashboard and select an auto-refresh interval (e.g., every 5 seconds, 1 minute, or 5 minutes). You can also adjust the time range for the data displayed in the dashboard by clicking the time picker in the top-right corner and choosing a predefined or custom time range.

13. **Configure Kibana alerts**: Create an alert in Kibana to notify you when a certain threshold of failed authentication attempts is reached within a specific time window. You can configure the alert to send notifications via email, Slack, PagerDuty, or other channels.

14. **Monitor and analyze unauthorized access attempts**: Regularly review the Kibana dashboards and alerts to identify patterns or trends in unauthorized access attempts. Use this information to improve your authentication and access control mechanisms, such as implementing multi-factor authentication, strengthening password policies, and tightening access controls.

By following these steps, you can effectively use the EFK stack to detect and respond to unauthorized access attempts in your Kubernetes environment. This will help you maintain a strong security posture and protect your infrastructure from potential breaches and attacks.

Detecting privilege escalation attempts

Privilege escalation is a common step in a pentest, and detecting privilege escalation attempts can help during the investigation stage and also in promptly taking action to protect the infrastructure in real time. Let us learn how to detect privilege escalation attempts using the EFK stack:

1. **Configure Fluentd to collect relevant logs**: Ensure Fluentd is configured to collect logs from all Kubernetes components, including nodes, Pods, containers, and the Kubernetes API server, with a special focus on logs related to role assignments and role changes. These logs will provide information about any attempts to escalate privileges within the cluster:

```
<source>
  @type tail
  path /var/log/containers/*.log
  pos_file /var/log/fluentd-containers.log.pos
  tag kubernetes.*
  read_from_head true
  <parse>
    @type json
    time_format %Y-%m-%dT%H:%M:%S.%NZ
  </parse>
</source>
<filter kubernetes.**>
@type kubernetes_metadata
</filter>
```

2. **Create Kibana visualizations and dashboards**: Log in to Kibana and create visualizations for monitoring privilege escalation attempts. The steps are similar to the unauthorized access attempts, but you will focus on visualizing data related to role changes, role assignments, and the usage of elevated privileges.

3. **Create an index pattern**: Define an index pattern in Kibana that matches the Elasticsearch indices where your privilege escalation logs are stored.

4. **Navigate to Visualize**: To create a new visualization, click on the **Visualize** tab in the left-hand menu. Click **Create visualization** to start creating a new visualization.

5. **Select a visualization type**: Choose a visualization type that best represents the data you want to display. To monitor privilege escalation attempts, a histogram or line chart might be suitable.

6. **Choose the index pattern**: Select the index pattern you created earlier (e.g., `fluentd-*`) as the data source for your visualization.

7. **Configure the visualization**: Now, you'll configure the visualization by defining the X-axis (horizontal) and Y-axis (vertical) data. For our example of monitoring privilege escalation attempts, follow these steps:

   I.   Under the **Metrics** section, click on **Y-axis** and select **Count** as the aggregation type. This will display the number of privilege escalation attempts on the Y-axis.

   II.   Under the **Buckets** section, click on **X-axis** and choose **Date Histogram** as the aggregation type. This will display the time on the X-axis. Choose the **@timestamp** field and an appropriate time interval (e.g., `1m` for one-minute intervals).

   III. To further group the data by user or service account, click **Add sub-buckets** under the **Buckets** section. Select **Split series** and choose **Terms** as the aggregation type. Then, select the field you want to group by, such as `user_name` or `service_account`. Set the **Order by** option to **Descending** and **Size** to a suitable value (e.g., 5 to display the top five users or service accounts).

8. **Save the visualization**: Once you've configured the visualization, click the **Save** button at the top of the page. Give your visualization a descriptive name (e.g., **Privilege Escalation Attempts**) and click **Save**.

9. **Create a dashboard**: Now that you've created a visualization, you can add it to a dashboard. Follow the same steps as outlined in the unauthorized access attempts tutorial.

10. **Customize the dashboard**: Dashboards in Kibana are highly customizable, allowing you to add multiple visualizations, adjust the layout, and apply filters to focus on specific data. You can add more visualizations to the dashboard to visualize other types of data, such as the specific roles or privileges that are being escalated, or the specific services or workloads where the escalation attempts are occurring.

11. **Apply filters**: Filters help you narrow down the data displayed in the dashboard. To apply a filter, click the **Add filter** button at the top of the dashboard. You can create filters based on specific fields (e.g., user, service account, role, or Kubernetes namespace) and conditions (e.g., equals, does not equal, is one of, or is not one of). For example, you can create a filter to display only privilege escalation attempts by a specific user or within a specific Kubernetes namespace.

12. **Set auto-refresh and time range**: To keep your dashboard up to date, you can configure it to refresh automatically at regular intervals. You can also adjust the time range for the data displayed in the dashboard by choosing a predefined or custom time range.

13. **Configure Kibana alerts**: Create an alert in Kibana to notify you when a certain threshold of privilege escalation attempts is reached within a specific time window. This can help you respond quickly to potential security incidents. You can configure the alert to send notifications via email, Slack, PagerDuty, or other channels.

14. **Monitor and analyze privilege escalation attempts**: Regularly review the Kibana dashboards and alerts to identify patterns or trends in privilege escalation attempts. Use this information to improve your RBAC policies, such as implementing least privilege principles, strengthening role assignment processes, and auditing role changes regularly.

Important note

Privilege escalation attempts can be a sign of insider threats or compromised accounts. If you detect repeated or successful privilege escalation attempts, it's crucial to investigate further to determine the cause and take appropriate action. This may involve incident response processes, forensic analysis, or even involving law enforcement if necessary.

If you notice, there is a pattern in detecting security threats using the EFK stack in the previous two examples, you would notice that the implementation part remains the same for most of it. Once it is deployed, however, the visualization methodology and the indexes that we examine change. Hence, as an exercise, I encourage you to try implementing a solution to detect security threats for the other examples, as shared previously.

Optimizing the EFK stack for Kubernetes

The EFK stack can be deployed in any environment for any host server; however, since we are mostly focusing on Kubernetes, let us look into a few tips and tricks to optimize the deployment for Kubernetes.

To improve the performance, reliability, and security of the EFK stack in a Kubernetes environment, consider implementing the following optimizations:

- Use Elasticsearch **index lifecycle management** (**ILM**) to manage the life cycle of your indices, optimizing storage usage and search performance

- Employ Elasticsearch index templates to define settings, mappings, and aliases for new indices, ensuring consistency and optimal configurations

- Implement Fluentd persistent queues to prevent data loss during Fluentd restarts or crashes

- Use Kibana spaces to organize and separate visualizations, dashboards, and other Kibana objects, simplifying management and improving access control

By following these best practices and optimizations, you'll be able to establish a comprehensive cloud security operations process, ensuring the security and resilience of your cloud-native applications. With the EFK stack in place, your team will have a powerful and flexible solution for centralized logging and monitoring in your Kubernetes environment.

Creating alerting and webhooks within different platforms

Alerting is a crucial part of monitoring your cloud-native infrastructure. It enables you to be proactive about addressing issues, ensuring system stability and reliability. In this section, we'll cover creating alerting rules in Prometheus and configuring webhook notifications for different platforms. We'll also look at how to automate incident response with custom scripts and tools.

Creating alerting rules in Prometheus

Prometheus is an open source monitoring and alerting toolkit designed specifically for microservices and containers. It collects metrics from configured targets at given intervals, evaluates rule expressions, and displays the results.

Alerting rules in Prometheus are divided into two parts. First, there are rule files, which contain the alerting rules themselves. Second, there is Alertmanager, which handles alerts sent by client applications such as the Prometheus server.

Prometheus alerting rules are essentially simple logic expressions that are evaluated every time Prometheus collects data. If the conditions defined by the alerting rule are satisfied, Prometheus sends an alert to Alertmanager.

Let's examine a few examples of alerting rules that could be useful to detect potential security threats:

1. **Excessive HTTP errors**: An unusually high number of HTTP errors could indicate an ongoing attack on your application, such as a DDoS attack, or it could be the result of an internal error:

```
- alert: HighHttpErrorRate
  expr: rate(http_requests_total{status_code=~"5.."}[5m]) /
rate(http_requests_total[5m]) > 0.05
  for: 10m
  labels:
    severity: critical
  annotations:
    summary: High HTTP error rate
    description: "{{ $labels.instance }} has a high HTTP error
rate: {{ $value }}% errors over the last 10 minutes."
```

2. **High memory usage**: A sudden spike in memory usage could indicate a memory leak in your application or a security breach where an attacker is consuming resources:

```
- alert: HighMemoryUsage
  expr: (node_memory_MemTotal_bytes - node_memory_MemAvailable_
bytes) / node_memory_MemTotal_bytes > 0.8
  for: 5m
  labels:
    severity: warning
  annotations:
    summary: High memory usage
    description: "{{ $labels.instance }} memory usage is
dangerously high: {{ $value }}% used."
```

3. **Pods restarting frequently**: If your Kubernetes pods are constantly crashing and restarting, it could indicate an issue with the application they're running or a potential security threat:

```
- alert: K8SPodFrequentRestart
  expr: increase(kube_pod_container_status_restarts_total[15m])
> 5
  for: 10m
  labels:
    severity: warning
  annotations:
    summary: Kubernetes Pod Restarting Frequently
    description: "Pod {{ $labels.namespace }}/{{ $labels.pod }}
is restarting frequently: {{ $value }} restarts in the last 15
minutes."
```

An example that comes to mind is a major e-commerce company that experienced frequent attacks in which its CPU usage spiked dramatically. A Prometheus alert rule that would have helped detect this threat earlier could be as follows:

```
- alert: HighCpuUsage
  expr: 100 - (avg by (instance) (irate(node_cpu_seconds_
total{mode="idle"}[5m])) * 100) > 80
  for: 10m
  labels:
    severity: critical
  annotations:
    summary: "Instance {{ $labels.instance }} CPU usage is dangerously
high"
    description: "{{ $labels.instance }} CPU usage is above 80%."
```

These are just a few examples of the type of alerts you can set up in Prometheus to monitor for potential security threats. By creating alerting rules based on your unique requirements and environment, you can ensure your systems are well monitored and any security threats are swiftly identified.

Configuring webhook notifications for different platforms (e.g., Slack)

Alertmanager handles alerts sent by the Prometheus server and takes care of deduplicating, grouping, and routing them to the correct receiver. It can send notifications via methods such as email, PagerDuty, or Opsgenie, and also to webhook endpoints.

Let's focus on configuring Alertmanager to send notifications to Slack:

1. **Define the Slack receiver in Alertmanager**: In your Alertmanager configuration, define a new receiver for Slack:

```
receivers:
- name: 'slack-notifications'
  slack_configs:
  - api_url: 'https://hooks.slack.com/services/<complete_URI>
    channel: '#alerts'
    send_resolved: true
```

In this configuration, `api_url` is your Slack webhook URL, `channel` is the Slack channel where you want to send the alerts, and `send_resolved` is set to `true`, which means that once the alert condition is no longer present, a resolution message will be sent to the Slack channel.

2. **Define a route for the alerts**: In the same configuration file, define a route that sends the alerts to the `slack-notifications` receiver:

```
route:
  group_by: ['alertname', 'cluster', 'service']
  group_wait: 30s
  group_interval: 5m
  repeat_interval: 3h
  receiver: 'slack-notifications'
  routes:
  - match:
      severity: critical
    receiver: 'slack-notifications'
```

In this configuration, `group_by` is used to group alerts with the same values for `alertname`, `cluster`, and `service`. The amount of time to wait before sending a notification about new alerts added to a group of alerts is `group_wait`. Meanwhile, `group_interval` is how long to wait before sending a notification about alerts that have been added to a group, and `repeat_interval` is how long to wait before resending a notification. The `slack-notifications` receiver will get the alerts that have a severity of `critical`.

3. **Reload or restart Alertmanager**: After you've made these changes, you'll need to reload or restart Alertmanager for the changes to take effect.

Now, whenever an alert is triggered, you'll receive a notification in the specified Slack channel.

Automating incident response with custom scripts and tools

Automating incident response can significantly reduce the time it takes to address security threats. This can be achieved by writing custom scripts or using incident response tools that integrate with your existing monitoring and alerting tools.

For instance, you can write a script that automatically isolates a compromised container or pod when a specific Prometheus alert is triggered. This script can be executed automatically using tools such as Ansible or Chef, or even from a custom webhook.

Another example is integrating your alerting system with an incident management platform such as Opsgenie or VictorOps. These platforms offer features such as on-call schedules, escalation policies, and automated remediation, which can help you respond to incidents more effectively.

Here's a basic example of how you might set up an automated response to a specific Prometheus alert:

1. **Create a custom script**: This script will be executed when the alert is triggered. For example, the script could use `kubectl` to isolate a compromised Kubernetes pod:

```bash
#!/bin/bash

# Get the name of the compromised pod from the alert payload
pod_name=$(jq -r .alerts[0].labels.pod_name $1)

# Use kubectl to isolate the pod
kubectl label pods $pod_name quarantine=true
```

2. **Configure the alert to trigger the script**: In the Alertmanager configuration, define a new receiver that triggers the custom script:

```
receivers:
- name: 'custom-script'
  webhook_configs:
  - url: 'http://localhost:5001/webhook'
```

In this configuration, `url` is the endpoint that triggers the custom script. This could be a simple Flask app that runs the script when it receives a `POST` request.

3. **Define a route for the alerts**: Define a route that sends the alerts to the `custom-script` receiver:

```
route:
  ...
  receiver: 'custom-script'
routes:
- match:
  severity: critical
  receiver: 'custom-script'
```

In this configuration, the `custom-script` receiver will get the alerts that have a severity of `critical`.

4. **Test the setup**: After setting up the custom script and updating the Alertmanager configuration, it's important to test the setup to ensure that the script is triggered correctly when the alert conditions are met.

This example demonstrates a simple automated response to a specific security threat. Depending on your requirements, you can create more complex scripts or use incident response tools to automate various tasks such as notifying the on-call team, creating a ticket in your issue tracking system, or even automatically scaling your infrastructure to handle an increase in load.

As you can see, automating incident response is a powerful way to reduce the time it takes to respond to security threats, and it can also help reduce human error. By integrating your monitoring and alerting tools with your incident response processes, you can create a robust security posture for your cloud-native environment.

Automated security lapse findings

In the cloud-native landscape, a different set of considerations comes into play when considering automated security lapse findings. Cloud-native applications are built and run on cloud-based platforms and services, leveraging the full benefits of cloud computing. They are usually composed of microservices, deployed in containers, and orchestrated by Kubernetes or other similar systems. In such an environment, the complexity, scale, and dynamism necessitate a different approach to security.

Security Orchestration, Automation, and Response (SOAR) platforms

The cyber threat landscape is continuously evolving, and it's becoming increasingly complex for security teams to manage the numerous alerts generated by multiple security tools. This is where **SOAR** platforms come into play. SOAR platforms aid in collecting security threat data from various sources, allowing teams to respond to low-level threats automatically and freeing up time to focus on complex and critical security incidents.

What is SOAR?

SOAR is a stack of compatible software programs that allow an organization to collect data about security threats from multiple sources and respond to low-level security events without human assistance. It was introduced by Gartner, the research and advisory company, as a direct response to the overwhelming number of alerts security teams were handling.

The main components of SOAR are as follows:

- **Security Orchestration and Automation (SOA)**: SOA integrates security tools and automates tasks that were traditionally performed manually. This reduces the time taken to respond to an incident and mitigates the risk of human error.

- **Security Incident Response Platforms (SIRPs)**: SIRPs provide a single platform to manage the incident response process. This includes tracking incidents, assigning tasks, and documenting actions taken.

- **Threat Intelligence Platforms (TIPs)**: TIPs provide information about the latest threats and uses this data to enhance the other components of SOAR.

In a cloud-native context, a SOAR platform can gather data from a wide range of sources, including container security tools, Kubernetes audit logs, cloud service logs, and network monitoring tools.

For example, if a container security tool detects suspicious activity in a container, it can send an alert to the SOAR platform. The platform can then automatically isolate the affected container, spin up a new one, and notify the security team.

Similarly, if a network monitoring tool detects a potential DDoS attack, the SOAR platform can automatically adjust the network rules to block the offending IP addresses and prevent the attack from impacting the application's performance.

This level of automation and orchestration is crucial in a cloud-native environment, where the scale and speed of operations can make it impossible for humans to respond quickly enough.

Benefits of SOAR

SOAR platforms offer numerous benefits, some of which include the following:

- **Reduced response time**: Automating responses to common threats significantly reduces the time it takes to react, limiting potential damage
- **Improved efficiency**: By reducing the manual tasks security teams need to perform, SOAR allows them to focus on more complex and strategic tasks
- **Standardized processes**: SOAR platforms help to standardize the incident response process, ensuring a consistent approach to managing security threats
- **Reduced alert fatigue**: By dealing with routine alerts automatically, SOAR platforms reduce the alert fatigue that can lead to missed or ignored threats

Now that we realize the need for a SOAR platform and the benefits of using one platform, let us look into some of the existing SOAR platforms available on the market today.

SOAR platforms on the market

There are several SOAR platforms available on the market today. Some popular ones include the following:

- **IBM Resilient**: IBM Resilient is a robust SOAR platform that offers a wide range of capabilities, including incident response, threat intelligence, and security automation
- **Splunk Phantom**: Phantom, acquired by Splunk, provides automation, orchestration, and response capabilities, allowing teams to automate tasks, orchestrate workflows, and manage cases
- **Rapid7 InsightConnect**: InsightConnect is an SOA platform that enables accelerated incident response
- **Siemplify**: Siemplify provides a holistic SOAR platform that combines security orchestration, automation, and response into one end-to-end solution

Due to the dynamic nature of cloud-native environments, the number of security events can be enormous. SOAR platforms help to manage this complexity by integrating various security tools and automating the response to low-level security threats.

For instance, when a new container is deployed in a cloud-native application, it's necessary to verify that it's secure. This involves checking the container image for vulnerabilities, verifying that it's configured correctly, and monitoring its behavior for signs of malicious activity. This is a complex task that can be significantly simplified with a SOAR platform.

SOAR platforms designed for cloud-native environments can integrate with tools such as container security platforms, **cloud security posture management (CSPM)** tools, **cloud workload protection platforms (CWPPs)**, and Kubernetes security solutions. This allows them to collect security data from across the cloud-native stack, from the infrastructure level up to the application level.

Moreover, these platforms can automate responses to common threats in a cloud-native context. For example, if a container is found to be running a vulnerable version of a software package, the SOAR platform could automatically isolate it, notify the responsible team, create a ticket in the issue tracking system, and initiate the process to update the container image.

Integrating security tools and automating workflows

To successfully implement a SOAR platform, it's crucial to integrate your existing security tools and automate workflows effectively. Let's delve into this process.

Integrating security tools

A successful SOAR platform heavily relies on the integration of various security tools. These could include **Security Information and Event Management (SIEM)** systems, **Intrusion Detection/Prevention Systems (IDSs/IPSs)**, TIPs, endpoint protection platforms, and many others.

The integration process typically involves configuring your security tools to send data to your SOAR platform and configuring the SOAR platform to ingest this data. This often includes specifying the data format, setting up API keys or other authentication methods, and defining which types of events should trigger alerts or automated actions.

Most SOAR platforms provide pre-built integrations with popular security tools, but it's also possible to develop custom integrations using APIs or scripting languages. The integration process will vary depending on the specific tools and SOAR platform you're using, so always refer to the documentation provided by the vendors.

In cloud-native environments, integrating security tools and automating workflows involve several cloud-specific considerations. Security tools must be able to work with cloud APIs, understand the specifics of cloud services, and be able to handle the scale and dynamism of cloud-native applications.

For example, a security tool might need to integrate with a cloud provider's API to collect logs or security events, or with a Kubernetes API to monitor the state of the cluster. It might also need to understand cloud-specific constructs such as IAM roles, security groups, or Kubernetes network policies.

A successful SOAR implementation in a cloud-native context relies heavily on the integration of various security tools. These could include container security tools (such as Aqua Security or Sysdig), Kubernetes-native security solutions (such as StackRox or Tigera), cloud security tools (such as Palo Alto Networks Prisma or Check Point CloudGuard), and network security tools (such as Cisco Stealthwatch or Juniper Networks Sky ATP).

Most SOAR platforms provide pre-built integrations with these and other popular security tools, but it's also possible to develop custom integrations using APIs or scripting languages. The integration process will vary depending on the specific tools and SOAR platform you're using, so always refer to the documentation provided by the vendors.

Automating workflows involves creating automated responses to cloud-specific security threats. This might include things such as an IAM role being granted excessive permissions, a security group allowing too much inbound traffic, or a Kubernetes pod running in privileged mode. These workflows need to be able to interact with cloud APIs to take corrective action, such as modifying an IAM policy, updating a security group, or deleting a Kubernetes pod.

Automating workflows

Once your security tools are integrated, the next step is to automate your workflows. This involves defining a series of steps that should be taken when a certain event or alert is detected.

For example, let's consider a simple workflow for responding to a detected malware infection:

1. The endpoint protection platform detects a malware infection on a user's device and sends an alert to the SOAR platform.

2. The SOAR platform automatically isolates the infected device from the network to prevent the malware from spreading.

3. The SOAR platform sends an email to the user informing them of the issue and instructing them to contact the IT department.

4. The SOAR platform creates a ticket in the IT service management system to track the incident.

This workflow can be defined in the SOAR platform using a visual editor or a scripting language, depending on the platform. It's important to thoroughly test your automated workflows in a controlled environment before deploying them in production.

Building and maintaining a security automation playbook

A security automation playbook is a document or tool that outlines the organization's standard procedures for detecting, investigating, and responding to various types of security incidents. It's an essential part of any SOAR platform, as it provides the basis for automated workflows.

Elements of a security automation playbook

A comprehensive security automation playbook should include the following elements:

- **Incident types**: Define the various types of security incidents that your organization might face, such as malware infections, phishing attacks, or data breaches.

- **Detection methods**: Describe how each type of incident is detected. This might involve SIEM systems, IDSs/IPSs, endpoint protection platforms, or manual reports from users.

- **Response procedures**: Outline the steps that should be taken in response to each type of incident. These steps will form the basis for your automated workflows in the SOAR platform.

- **Roles and responsibilities**: Specify who is responsible for each step in the response procedure. This might include security analysts, IT staff, or management.

- **Communication procedures**: Define how information about incidents should be communicated within the organization. This includes who should be notified, what information should be shared, and how it should be communicated.

Now that we understand the modular concepts of security automation, let us look into how we can create a playbook, to avoid any repetition in dealing with security incidents.

Building a security automation playbook

Building a security automation playbook requires a thorough understanding of your organization's security landscape, including the threats you face, the tools you use to detect and respond to these threats, and the roles and responsibilities of your security team.

Start by conducting a risk assessment to identify the types of incidents that are most likely to occur and would have the most significant impact on your organization. Then, for each type of incident, outline the detection methods and response procedures.

When defining response procedures, consider both technical actions (e.g., isolating an infected device or blocking a malicious IP address) and non-technical actions (e.g., notifying affected users or reporting the incident to management). Remember that not all steps can or should be automated, especially when it comes to decision-making or actions that have a significant impact.

A security automation playbook for cloud-native applications requires understanding the unique security challenges that these environments pose. The playbook should cover the different types of threats that are relevant in a cloud-native context, such as misconfigured cloud resources, compromised cloud credentials, insecure container images, or vulnerabilities in Kubernetes.

The playbook should also define how to respond to these threats in an automated way. This might involve interacting with cloud APIs, using cloud-native security tools, or integrating with the cloud-native stack. For example, a response to a compromised cloud credential might involve disabling the credential via the cloud provider's API, using a cloud-native security tool to investigate the scope of the breach, and integrating with the cloud-native logging and monitoring stack to gather relevant data.

Security in the cloud-native landscape adds additional layers of complexity, but also offers unique opportunities for automation and orchestration. The dynamic, distributed nature of cloud-native applications, which are typically built as microservices running in containers and orchestrated by platforms such as Kubernetes, requires a new approach to security.

Let's consider a playbook that defines a workflow for responding to a detected container vulnerability:

1. The container security tool detects a vulnerability in a running container and sends an alert to the SOAR platform.

2. The SOAR platform automatically pulls the image of the vulnerable container for further analysis.

3. The SOAR platform creates a ticket in the IT service management system to track the incident and notifies the security team.

4. The security team investigates the vulnerability and decides on the appropriate response, which could include patching the container image, changing the configuration, or updating the application code.

5. The SOAR platform automatically deploys the updated container and verifies that the vulnerability has been addressed.

This playbook ensures a quick and consistent response to container vulnerabilities, reducing the window of opportunity for an attacker to exploit them.

Maintaining a security automation playbook

A security automation playbook is not a one-time document. It should be regularly updated to reflect changes in your organization's security landscape. This includes new threats, changes in your security tools or processes, and lessons learned from previous incidents.

Regularly review and update your playbook and consider conducting drills or simulations to test your procedures and ensure that everyone knows their roles and responsibilities. After each incident, conduct a post-mortem to identify any gaps or improvements that can be made, and update the playbook accordingly.

Case study – automating response to phishing attacks

Let's consider a practical example to illustrate how SOAR platforms, security tools integration, automated workflows, and security automation playbook can work together to automate the response to a common security incident: phishing attacks.

Phishing attacks are a common security threat that many organizations face. They involve malicious actors sending fraudulent emails pretending to be from reputable companies to induce individuals to reveal personal information, such as passwords and credit card numbers.

In our case, the organization uses a SOAR platform that is integrated with an email security gateway (for detecting phishing emails), an endpoint protection platform (for detecting malicious activity on user devices), and an IT service management system (for managing incidents).

Detection

The email security gateway is configured to detect potential phishing emails based on various indicators, such as the sender's reputation, the presence of suspicious links or attachments, and the use of social engineering techniques. When a potential phishing email is detected, the gateway blocks the email and sends an alert to the SOAR platform.

Automated response

The SOAR platform receives the alert and triggers an automated workflow for responding to phishing attacks. This workflow includes the following steps:

1. The SOAR platform sends an email to the intended recipient of the phishing email, informing them that a phishing attempt has been detected and blocked. The email includes information about phishing attacks and how to avoid them.

2. The SOAR platform creates a ticket in the IT service management system to track the incident. The ticket includes all the relevant information about the phishing attempt, such as the sender's email address, the content of the email, and the intended recipient.

3. The SOAR platform updates a dashboard that provides real-time information about security incidents, including the number and types of incidents detected, the status of the response, and any trends or patterns.

Security automation playbook

The organization's security automation playbook includes a section on phishing attacks. This section defines the detection methods (i.e., the email security gateway), the automated response procedures (i.e., the steps defined in the SOAR platform), and the roles and responsibilities (i.e., the SOAR platform manages the automated response, the IT department manages the IT service management system, and the security team reviews and updates the dashboard).

The playbook is reviewed and updated regularly to ensure that it accurately reflects the organization's procedures and threats. For example, if the organization starts using a new tool to detect phishing emails, the playbook is updated to include this tool in the detection methods.

This case study illustrates how SOAR platforms, security tools integration, automated workflows, and security automation playbooks can work together to automate the response to security incidents, improving efficiency and reducing the risk of human error.

Continuous monitoring and improvement

An essential part of maintaining a security automation playbook is continuous monitoring and improvement. The organization continuously monitors the effectiveness of the automated workflows and the accuracy of the playbook. They use metrics such as the number of incidents detected, the number of false positives, the time taken to respond to incidents, and the effectiveness of the response.

They also conduct regular reviews of the playbook to ensure that it is up to date and reflects the current threat landscape. They consider new threats and attack vectors, changes in the organization's systems and processes, and feedback from users and security staff.

For example, if the organization noticed an increase in phishing attacks impersonating a specific company, it might update the playbook to include additional steps for detecting and responding to these attacks. They might also conduct a training session for employees to make them aware of this new threat.

Automating regular security audits

In addition to automating the response to security incidents, the organization also uses automation to conduct regular security audits. They have a separate automated workflow in the SOAR platform that checks the configuration of their systems against a list of best practices, checks for outdated software versions, and checks for any signs of compromise.

This automated audit runs on a regular schedule and sends a report to the security team. If any issues are detected, the SOAR platform automatically creates a ticket in the IT service management system.

This approach allows the organization to proactively detect and address potential security issues before they can be exploited by malicious actors. It also reduces the manual effort involved in conducting security audits.

In conclusion, automated security lapse findings are a critical component of a robust and proactive security strategy. By leveraging SOAR platforms, integrating security tools, automating workflows, and maintaining a security automation playbook, organizations can respond to security incidents more quickly and efficiently, reducing the risk of damage or data loss.

However, automation is not a silver bullet. It is essential to have skilled security professionals who can interpret the results of automated systems, handle complex incidents, and continuously monitor and improve the organization's security posture. Automation is a tool that can help these professionals do their job more effectively, not a replacement for their expertise and judgment.

Remember, the ultimate goal is not to respond to security incidents but to prevent them from happening in the first place. By combining automation with a strong security culture, continuous learning, and proactive threat hunting, organizations can significantly reduce their risk of security lapses.

Summary

In this chapter, we embarked on an enlightening journey, diving deep into the realms of centralized logging, automated alerting, and effective security orchestration for cloud-native applications.

We began with the exploration of the EFK stack – Elasticsearch, Fluentd, and Kibana – in the context of a Kubernetes environment. Elasticsearch serves as the search and analytics engine, Fluentd is for data collection and aggregation, and Kibana is for data visualization. We provided a detailed, step-by-step guide on setting up and securing an EFK stack on a Kubernetes cluster. We discussed the importance of PersistentVolumes to Elasticsearch data and the role of Helm, a package manager for Kubernetes, to ease the installation process. We also covered the critical aspects of securing the EFK stack for a production environment, and its maintenance and monitoring.

Following the EFK stack, we moved on to automated alerting systems, shedding light on the role of tools such as Prometheus. We discussed how Prometheus can be used to monitor Kubernetes clusters, create alerting rules, and send alerts to alert managers. We emphasized the role of such systems in promptly notifying the relevant teams about any potential issues, thereby preventing minor problems from escalating into major crises.

Next, we delved into defining runbooks and playbooks for efficient incident management. We stressed the importance of detailed, clear, and concise documentation that outlines procedures to mitigate known issues. We also touched upon the role of automation in executing these runbooks and playbooks, reducing human error and response time.

In the latter part of the chapter, we explored SOAR platforms, focusing on their role in automating responses to security threats, thereby allowing security teams to focus on more sophisticated and critical security incidents.

We wrapped up this chapter with a case study on phishing attacks, illustrating the practical application of all the concepts discussed in this chapter, from the EFK stack to SOAR platforms and automated workflows.

As we proceed to the next chapter, we will delve deeper into the world of DevSecOps, a philosophy that integrates security practices within the DevOps process. The next chapter is a natural progression from our discussion on securing the EFK stack and implementing automated alerting and response systems. We will explore how to automate security checks within the CI/CD pipeline and implement security in the code and infrastructure, thus making it a continuation of this chapter's themes.

Quiz

Answer the following questions to test your knowledge of this chapter:

- Describe the role of each component in the EFK stack and how they interact within a Kubernetes environment. Why is each component crucial, and what unique function does it bring to the table? Please provide examples to substantiate your answer.

- Discuss the importance of automated alerting in a production environment. How does Prometheus aid in this process? Elaborate on how you might set up a basic alerting rule in Prometheus, citing an example scenario.

- What is the purpose of runbooks and playbooks in incident management? Describe a situation where they can be crucial and how automation can enhance their execution.

- Explain the concept of SOAR platforms, and how they can streamline the response to security threats. Can you outline a scenario where a SOAR platform might significantly reduce the time and effort spent on a security incident?

- Reflecting on the phishing attack case study, how do the concepts discussed in this chapter – from the EFK stack, automated alerting, runbooks, and playbooks to SOAR platforms – come together to address such a situation? How would you create an automated detection and alerting system for other types of attacks?

Further readings

- https://www.elastic.co/what-is/elk-stack

- https://prometheus.io/docs/introduction/overview/

- https://sysdig.com/blog/kubernetes-monitoring-prometheus/

- https://grafana.com/tutorials/grafana-fundamentals/

- https://www.cyberark.com/resources/blog/securing-kubernetes-clusters-by-eliminating-risky-permissions

- https://www.gartner.com/en/documents/3899465/security-orchestration-automation-and-response-soar-t

- https://www.splunk.com/en_us/software/siem-security-orchestration.html

- https://developers.redhat.com/blog/2019/02/21/pods-and-privileges/

- https://kubernetes.io/docs/concepts/configuration/secret/

- https://www.elastic.co/guide/en/elasticsearch/reference/current/docker.html

- https://kubernetes.io/docs/tasks/configure-pod-container/security-context/

- https://docs.fluentd.org/

- https://www.aquasec.com/wiki/display/containers/ Kubernetes+Audit+Logs

- https://www.elastic.co/guide/en/kibana/current/dashboard.html

- https://www.datadoghq.com/blog/monitoring-kubernetes-performance-metrics/

8

DevSecOps Practices for Cloud Native

As organizations shift toward cloud native applications, the need for robust security practices is more important than ever. This is where DevSecOps, a philosophy that integrates security practices within the DevOps process, plays a crucial role. In this chapter, we will delve into the various aspects of DevSecOps, focusing on **Infrastructure as Code (IaC)**, **Policy as Code (PaC)**, and **continuous integration/continuous deployment (CI/CD)** platforms. This chapter will teach you how to automate most of the processes you learned in the previous chapters.

By the end of this chapter, you will have a comprehensive understanding of these concepts and the open source tools that aid in implementing DevSecOps practices. You will learn how to secure the pipeline and code development using these tools, and how to detect security misconfigurations in IaC scripts. This knowledge will empower you to build more secure, efficient, and resilient cloud-native applications.

The lessons in this chapter are not only informative but also practical. You will be introduced to real-world scenarios and hands-on activities that will help you apply what you've learned. This practical approach will equip you with the skills needed to navigate the complex landscape of cloud-native application security.

In this chapter, we're going to cover the following main topics:

- Infrastructure as Code
- Policy as Code
- CI/CD platforms
- Security tools for pipeline and code development
- Tools for detecting security misconfigurations in IaC scripts

Technical requirements

The following are the technical requirements for this chapter:

- Terraform v1.4.6
- HashiCorp Vault v1.13.2
- Checkov v2.3.245
- OPA v0.52.0
- GitHub Actions
- AWS

Infrastructure as Code

DevSecOps, a portmanteau of **Development**, **Security**, and **Operations**, is a philosophy that integrates security practices within the DevOps process. It aims to embed security in every part of the development process. DevSecOps involves continuous integration, continuous deployment, IaC, PaC, security tools, and detection of security misconfigurations.

Here's a diagram that illustrates the concept of DevSecOps:

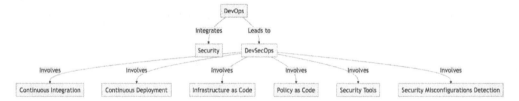

Figure 8.1 – DevSecOps components tree

In traditional development models, security was often an afterthought, typically addressed at the end of the development cycle. This approach often led to vulnerabilities being discovered late in the process, making them costly and time-consuming to fix. DevSecOps addresses this issue by integrating security practices right from the initial stages of the development cycle. This shift-left approach to security ensures that vulnerabilities are identified and mitigated early, reducing the overall risk.

DevSecOps encourages a culture of shared responsibility for security. In this model, everyone involved in the application development process is responsible for security. This collaborative approach not only improves the security posture of the applications but also accelerates the development process as security issues are addressed in real time.

CI and CD are key components of DevSecOps. CI involves merging all developers' working copies to a shared mainline several times a day. This practice helps detect integration errors as quickly as possible. CD, on the other hand, involves the automated deployment of code to production, ensuring that the software can be reliably released at any time.

IaC is another important aspect of DevSecOps. IaC is the process of managing and provisioning computing infrastructure with machine-readable definition files, rather than physical hardware configuration or interactive configuration tools. This practice brings a level of efficiency and repeatability to the provisioning and deployment process that was previously difficult to achieve.

PaC is a practice where policy definitions are written in code and managed as code files in version control. This allows policies to be versioned, tested, and applied consistently across the infrastructure.

Security tools play a crucial role in the DevSecOps process. These tools can help detect vulnerabilities, enforce security policies, and provide visibility into the security posture of the applications.

Finally, the detection of security misconfigurations in IaC scripts is an important part of DevSecOps. Tools such as Checkov and Terrascan can identify and mitigate security risks in IaC scripts, enhancing the overall security of the applications.

By integrating these practices, DevSecOps provides a robust framework for securing cloud-native applications. It ensures that security is not just the responsibility of a single team but is a shared responsibility across all teams involved in the application development process. Now, let's look at how IaC fits into the larger DevSecOps picture by learning the importance of DevSecOps in creating the need for IaC.

The importance of DevSecOps

In the rapidly evolving world of software development, security can no longer be an afterthought or a separate phase in the development life cycle. With the increasing complexity of applications and the rising sophistication of cyber threats, security needs to be integrated into every step of the development process. This is where DevSecOps comes in.

DevSecOps is not just a set of tools or practices, but a cultural shift. It represents a move away from the traditional siloed approach where the development, security, and operations teams work separately. Instead, DevSecOps promotes a collaborative approach where all teams work together and share the responsibility for security. This ensures that security considerations are not overlooked and that they are addressed as early as possible in the development process.

By integrating security into the DevOps process, DevSecOps helps reduce the risk of security issues and vulnerabilities. It allows for continuous security monitoring and automated testing, which can help you detect and fix issues more quickly and efficiently. This can lead to improved security, faster development times, and lower costs.

DevSecOps in practice

In practice, implementing DevSecOps involves a combination of cultural changes, process changes, and the use of specific tools and technologies. Here are some key elements of a successful DevSecOps implementation:

- **Cultural changes**: DevSecOps requires a shift in mindset where all team members take responsibility for security. This involves fostering a culture of collaboration and transparency and promoting continuous learning and improvement.

- **Process changes**: DevSecOps involves integrating security practices into every stage of the DevOps process. This includes practices such as threat modeling in the design phase, code reviews and static code analysis in the development phase, and continuous monitoring and incident response in the deployment and operation phases.

- **Tools and technologies**: Many tools and technologies support a DevSecOps approach. These include tools for CI/CD, IaC, automated testing, security scanning, and monitoring.

By implementing DevSecOps, organizations can create a more secure and efficient development process and build applications that are more resilient to cyber threats. However, it's important to remember that DevSecOps is not a one-size-fits-all solution, and each organization will need to adapt the approach to fit its specific needs and context.

Continuous integration and continuous deployment (CI/CD) in DevSecOps

CI and CD are fundamental to the DevSecOps approach. They represent practices that aim to automate the processes of software delivery and infrastructure changes.

CI is a development practice where developers integrate code into a shared repository frequently, usually multiple times per day. Each integration can then be verified by an automated build and automated tests. The primary goal of CI is to prevent integration problems, which were quite common when developers worked in isolation for an extended period. CI not only reduces integration issues but also enables rapid application development.

CD is closely related to CI and refers to the practice of automating the delivery of software to production. With CD, every change that passes all stages of your production pipeline is released to your customers. There's no human intervention, and only a failed test will prevent a new change from being deployed to production.

CI/CD pipelines are designed to be a part of the DevSecOps strategy, which integrates developers and operations staff throughout the entire life cycle of an application, from design through the development process to production support. CI/CD pipelines help reduce manual errors, provide standardized development feedback loops, and enable fast product iterations.

Infrastructure as Code (IaC) and Policy as Code in DevSecOps

IaC involves managing and provisioning infrastructure through machine-readable definition files, rather than physical hardware configuration or interactive configuration tools. This allows for a simple, repeatable, and consistent environment. In a DevSecOps context, IaC can be used to automate the setup of environments and ensure they are set up consistently and with security controls in place.

PaC is the concept of writing code in high-level languages to manage and automate policies. These policies can be used to enforce certain standards or behaviors in the development and deployment process. In a DevSecOps context, PaC can be used to ensure that security policies are consistently applied across all environments.

Both IaC and PaC are important practices in DevSecOps as they help automate and standardize processes, reduce the risk of human error, and ensure that security is consistently enforced.

Security tools in DevSecOps

Many security tools can support a DevSecOps approach. These tools can be used to automate security tasks, monitor security risks, and help teams respond to security incidents more effectively.

Some of these tools include the following:

- **Static application security testing (SAST)** tools, which can analyze source code to find security vulnerabilities
- **Dynamic application security testing (DAST)** tools, which can find vulnerabilities in a running application
- **Security information and event management (SIEM)** tools, which can collect and analyze security events and logs
- Incident response tools, which can help teams respond to security incidents more effectively

These tools can be integrated into the CI/CD pipeline to provide continuous security feedback and ensure that security is considered at every stage of the development process.

Let's visualize the role of CI, CD, IaC, PaC, and security tools in the DevSecOps approach with the help of a diagram:

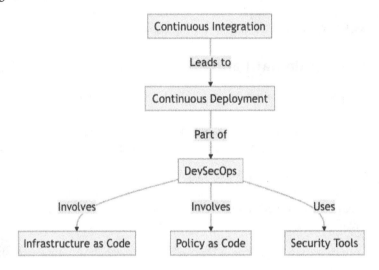

Figure 8.2 – DevSecOps components in CI/CD

As depicted in this diagram, CI leads to CD, both of which form an integral part of the DevSecOps approach. DevSecOps also involves the use of IaC and PaC. These practices help automate and standardize processes, reduce the risk of human error, and ensure that security is consistently enforced. Additionally, DevSecOps makes use of various security tools to automate security tasks, monitor security risks, and help teams respond to security incidents more effectively.

By integrating these practices and tools, DevSecOps provides a robust framework for securing cloud-native applications. It ensures that security is not just the responsibility of a single team but is a shared responsibility across all teams involved in the application development process.

IaC is a key practice in DevSecOps that involves managing and provisioning computer data centers through machine-readable definition files, rather than physical hardware configuration or interactive configuration tools. IaC brings a level of efficiency and repeatability to the provisioning and deployment process that was previously difficult to achieve.

Security implications of IaC

From a security perspective, IaC can significantly enhance the security posture of an organization. By defining infrastructure in code, you can apply the same security-focused practices that you use for application code to your infrastructure. This includes version control, code review, automated testing, and CI/CD.

IaC allows you to consistently apply security controls to infrastructure configurations, reducing the likelihood of misconfigurations that could lead to security vulnerabilities. It also allows you to audit and track infrastructure changes, which can help in identifying and investigating potential security incidents.

In this chapter, we will be focusing on a famous use case for automating the process of provisioning cloud-native infrastructure – Hashicorp Terraform.

Terraform – a comprehensive overview

Terraform, an open source tool developed by HashiCorp, has emerged as a key player in the IaC landscape. It allows developers and operators to provision and manage their infrastructure using a high-level configuration language known as **HashiCorp Configuration Language** (**HCL**). Terraform's platform-agnostic approach means it can be used with a wide array of service providers, making it a popular choice for multi-cloud deployments.

Terraform's power lies in its declarative nature. Instead of writing scripts to describe the procedural steps your infrastructure setup requires, you describe in HCL what your infrastructure should look like and Terraform takes care of realizing this state. This approach minimizes human error and increases reproducibility as the same code can be used to spin up identical environments.

The importance of security in Terraform scripts

While Terraform can help streamline your infrastructure management, it's not without its potential pitfalls. One of the most significant of these is the risk of security misconfigurations.

Security misconfigurations in Terraform scripts can lead to serious vulnerabilities, exposing your infrastructure to potential attacks. These misconfigurations can occur in several ways, including but not limited to the following:

- **Hardcoded secrets**: Embedding plaintext secrets such as API keys, passwords, or certificates directly in your Terraform scripts can expose these sensitive details, especially when scripts are stored in version control systems or shared among teams.

- **Overly permissive access controls**: Defining overly broad access rules in your Terraform scripts can leave your resources open to unauthorized access. This could be in the form of security groups that allow traffic from any IP address, or IAM policies that grant excessive permissions.

- **Publicly accessible resources**: Making cloud resources such as storage buckets or databases publicly accessible can lead to data leakage or unauthorized data manipulation.

- **Unencrypted data**: Failing to enable encryption for data at rest or in transit can make your data vulnerable to interception.

Learning about Terraform security misconfigurations

Understanding and learning about these potential security misconfigurations in Terraform is crucial for several reasons:

- **Preventing data breaches**: Misconfigurations can lead to data breaches, which can be costly not only in terms of financial loss but also damage to a company's reputation.

- **Regulatory compliance**: Many industries are subject to regulations that require certain security controls to be in place. Understanding security configurations in Terraform can help ensure compliance with these regulations.

- **Security best practices**: As part of a DevSecOps approach, it's important to incorporate security best practices into all stages of your development and operations processes. This includes infrastructure management with Terraform.

In the following sections, we'll delve deeper into each of these potential security misconfigurations, providing real-world examples and explaining how to avoid them. We'll also introduce tools that can help automate the detection of these issues, helping you ensure your Terraform scripts are as secure as possible.

Hardcoded secrets

Hardcoded secrets are a common security misconfiguration in Terraform scripts. This occurs when sensitive information, such as API keys, passwords, or certificates, are embedded directly in the Terraform scripts. This is problematic for several reasons:

- **Exposure of sensitive information**: If your Terraform scripts are stored in a version control system, anyone with access to the repository can view these secrets. This could include other team members who don't need access to these secrets, or even external actors if the repository is public.

- **Lack of rotation**: Hardcoded secrets are often not rotated regularly. If an attacker gains access to a secret, they can continue to use it so long as it remains valid.

- **Non-unique secrets**: Hardcoded secrets are often reused across different scripts or environments. This means that if one script or environment is compromised, others may be at risk as well.

To mitigate the risks associated with hardcoded secrets, you should do the following:

- **Use a secret management tool**: Tools such as HashiCorp's Vault, AWS Secrets Manager, or Azure Key Vault can securely store secrets and provide them to your Terraform scripts as needed. We will learn about this in more detail in the next few sections.

- **Rotate secrets regularly**: Regularly rotating your secrets reduces the window of opportunity for an attacker if they do gain access to a secret.

- **Use unique secrets**: Using unique secrets for each script or environment ensures that a compromise in one area doesn't put others at risk.

Using Vault with Terraform for secrets management

Let's see how we can use Vault to provide a secret to a Terraform script:

1. Install Vault on your local machine or your Kubernetes cluster (whatever environment you are using for importing secrets).

2. Start a Vault server in dev mode by running the following command:

   ```
   $ vault server -dev
   ```

3. In a new Terminal window, set the VAULT_ADDR environment variable so that it points to your Vault server by running the following command:

   ```
   $ export VAULT_ADDR='http://127.0.0.1:8200'
   ```

4. Write a secret to Vault by running the following command:

   ```
   $ vault kv put secret/mysecret value=mysecretvalue
   ```

5. Create a file named main.tf with the following content:

   ```
   provider "vault" {}

   data "vault_generic_secret" "mysecret" {
     path = "secret/mysecret"
   }

   output "mysecret" {
     value = data.vault_generic_secret.mysecret.data["value"]
   }
   ```

6. Run the following command to initialize your Terraform configuration:

   ```
   $ terraform init
   ```

7. Run the following command to read the secret from Vault and output it:

   ```
   $ terraform apply
   ```

This script uses the Vault provider for Terraform to read a secret from Vault and output it. Note that the secret is not hardcoded in the Terraform script.

In the next few sections, we'll explore other common Terraform security misconfigurations and how to mitigate them. We'll also provide more hands-on tutorials to help you secure your Terraform scripts.

Overly permissive access controls

Another common security misconfiguration in Terraform scripts is overly permissive access controls. This occurs when the access rules defined in your Terraform scripts allow more access than necessary – for example, a security group in AWS that allows traffic from any IP address or an IAM policy that grants excessive permissions.

Overly permissive access controls can lead to unauthorized access to your resources. An attacker could exploit these permissions to gain access to your resources, potentially leading to data breaches or other security incidents.

To mitigate the risks associated with overly permissive access controls, you should do the following:

- **Follow the principle of least privilege**: This principle states that a user or system should have the minimum permissions necessary to perform their tasks. In the context of Terraform, this means that your scripts should grant the least amount of access necessary for your resources to function correctly.

- **Regularly review and update access controls**: Over time, your access control needs may change. Regularly reviewing and updating your access controls can help ensure that they remain appropriate for your needs.

- **Use Infrastructure as Code to manage access controls**: By managing your access controls with Terraform, you can take advantage of version control, code review, and automated testing to help ensure that your access controls are correct and secure.

Securing access controls with Terraform

Let's create a secure AWS security group with Terraform:

1. Install Terraform on your local machine.
2. Create a file named `main.tf` with the following content:

```
provider "aws" {
  region = "us-west-2"
}
resource "aws_security_group" "example" {
  name        = "example"
  description = "Example security group"
  ingress {
```

```
    from_port   = 22
    to_port     = 22
    protocol    = "tcp"
    cidr_blocks = ["your.ip.address/32"]
  }
}
```

3. Run `terraform init` to initialize your Terraform configuration.

4. Run `terraform apply` to create the security group. Terraform will show you a plan of what it's going to do and will prompt you for confirmation.

This script creates a security group that only allows SSH access (port 22) from your IP address. This is much more secure than allowing access from any IP address. You can now apply this script for all of your infrastructure provisions where you would want to implement ingress/egress security.

Publicly accessible resources

Publicly accessible resources are another common security misconfiguration in Terraform scripts. This occurs when resources such as storage buckets or databases are made publicly accessible. Publicly accessible resources can be accessed or modified by anyone, which can lead to data leakage or unauthorized data manipulation.

To mitigate the risks associated with publicly accessible resources, you should do the following:

*   **Limit public access**: Unless necessary, resources should not be publicly accessible. Access should be limited to only those who need it.

*   **Use access controls**: Access controls can be used to limit who can access a resource. This could be in the form of security groups, IAM policies, or bucket policies.

*   **Monitor access logs**: Access logs can provide insight into who is accessing your resources. Regularly monitoring these logs can help you detect unauthorized access.

Creating a secure AWS S3 bucket with Terraform

Let's create a secure AWS S3 bucket with Terraform:

1. Install Terraform on your local machine.

2. Create a file named `main.tf` with the following content:

```
provider "aws" {
  region = "us-west-2"
}

resource "aws_s3_bucket" "example" {
  bucket = "example-bucket"
```

```
acl    = "private"

tags = {
  Name          = "My private bucket"
  Environment = "Dev"
}
}
```

3. Run `terraform init` to initialize your Terraform configuration.

4. Run `terraform apply` to create the S3 bucket. Terraform will show you a plan of what it's going to do and will prompt you for confirmation.

This script creates an S3 bucket that is private (not publicly accessible) and tags it with the name and environment.

Encrypting data in an AWS S3 bucket with Terraform

The lack of encryption for data at rest or in transit is another common security misconfiguration in Terraform scripts. Encryption is a method of preventing unauthorized access to data by converting it into a format that is unreadable without the decryption key. If your data is not encrypted, it can be read by anyone who gains access to it.

To mitigate the risks associated with unencrypted data, you should do the following:

- **Encrypt data at rest**: Data at rest refers to data that is stored on physical or virtual disk drives, databases, or other storage media. This data should be encrypted to protect it if the storage medium is physically stolen or if an unauthorized person gains access to the data.

- **Encrypt data in transit**: Data in transit refers to data that is being transferred over a network. This data should be encrypted to protect it from being intercepted during transmission.

- **Use encryption features provided by your cloud provider**: Most cloud providers offer features to help you encrypt your data at rest and in transit. For example, AWS offers AWS **Key Management Service** (**KMS**) for managing encryption keys, and S3 buckets can be configured to automatically encrypt all data stored in them.

Let's create an AWS S3 bucket with Terraform and configure it to automatically encrypt all data stored in it:

1. Install Terraform on your local machine.

2. Create a file named `main.tf` with the following content:

```
provider "aws" {
  region = "us-west-2"
}
```

```
resource "aws_s3_bucket" "example" {
  bucket = "example-bucket"

  server_side_encryption_configuration {
    rule {
      apply_server_side_encryption_by_default {
        sse_algorithm = "AES256"
      }
    }
  }

  tags = {
    Name        = "My encrypted bucket"
    Environment = "Dev"
  }
}
```

3. Run `terraform init` to initialize your Terraform configuration.

4. Run `terraform apply` to create the S3 bucket. Terraform will show you a plan of what it's going to do and will prompt you for confirmation.

This script creates an S3 bucket and configures it to automatically encrypt all data stored in it using the AES256 encryption algorithm.

While there are many such case scenarios to explore, the list is endless, and a security review should be conducted, ideally, for provisioning all the infrastructure. Hence, it becomes very important to also use automated tools in the CI/CD pipelines that can automate the security scanning process for security misconfigurations. We are going to be looking at one such tool in this chapter – Checkov.

Checkov – a comprehensive overview

Checkov is an open source static code analysis tool for IaC. Developed by Bridgecrew, it's designed to scan cloud infrastructure managed in Terraform, CloudFormation, Kubernetes, ARM templates, and Dockerfiles to detect security and compliance misconfigurations.

Checkov operates on the principle of *shift left*, a practice in DevOps that involves introducing security earlier in the life cycle of application development. By scanning IaC scripts before they're deployed, Checkov can identify potential security issues before they become a part of your live infrastructure.

How Checkov works

Checkov uses a graph-based approach to evaluate the relationships between cloud resources and identify misconfigurations. It comes with over 500 built-in policies that cover a wide range of security best practices, compliance guidelines, and common misconfigurations.

When you run Checkov against your IaC scripts, it parses the files into a graph or a tree structure that represents the resources to be deployed and their configuration. Then, it evaluates this structure against its built-in policies to identify any violations.

Tools such as Checkov play a crucial role in maintaining secure and compliant cloud infrastructure. Here's why:

- **Early detection of security issues**: By scanning IaC scripts before they're deployed, Checkov can identify potential security issues before they become a part of your live infrastructure. This allows you to fix these issues before they can be exploited.

- **Compliance**: Checkov comes with built-in policies that align with common compliance frameworks, such as CIS Benchmarks, GDPR, HIPAA, and PCI-DSS. This makes it easier to ensure that your infrastructure is compliant with these frameworks.

- **Automation**: Checkov can be integrated into your CI/CD pipeline, allowing you to automate the process of checking your IaC scripts for security and compliance issues.

- **Developer education**: By providing immediate feedback on security and compliance issues, Checkov can help educate developers about these issues and how to avoid them.

Using Checkov in a live cloud-native environment

Let's learn how we can use Checkov to scan a Terraform script in a live cloud-native environment:

1. Install Checkov on your local machine.

2. Create a file named `main.tf` with the following content:

```
provider "aws" {
  region = "us-west-2"
}

resource "aws_s3_bucket" "example" {
  bucket = "example-bucket"
  acl    = "public-read"

  tags = {
    Name        = "My public bucket"
```

```
    Environment = "Dev"
  }
}
```

3. Run the following command to scan the Terraform script with Checkov:

    ```
    $ checkov -f main.tf
    ```

Checkov will report that the S3 bucket is publicly readable, which is a security risk.

To fix this vulnerability, you should change the `acl` line to `acl = "private"`, which will make the bucket private.

Deep diving into Checkov's features

Checkov offers a range of features that make it a powerful tool for securing your infrastructure as code. Let's delve deeper into some of these features:

- **Built-in policies**: Checkov comes with over 500 built-in policies that cover a wide range of security best practices, compliance guidelines, and common misconfigurations. These policies are regularly updated to reflect the latest security recommendations.

- **Custom policies**: In addition to its built-in policies, Checkov allows you to define custom policies. This is useful if you have specific security requirements that aren't covered by the built-in policies.

- **Policy as Code**: Checkov uses a PaC approach, which means that its policies are defined in code. This allows you to version control your policies, track changes over time, and collaborate with your team on policy development.

- **Multi-framework support**: Checkov supports a range of IaC frameworks, including Terraform, CloudFormation, Kubernetes, ARM templates, and Dockerfiles. This means you can use Checkov to secure your infrastructure regardless of which IaC framework you're using.

- **CI/CD integration**: Checkov can be integrated into your CI/CD pipeline, allowing you to automate the process of checking your IaC scripts for security and compliance issues.

Creating a custom policy in Checkov

Let's learn how we can create a custom policy in Checkov:

1. Create a file named `checkov.yml` in the root of your project with the following content:

    ```
    checkov:
      checks:
        - name: "Ensure S3 buckets are not public"
          id: "CKV_CUSTOM_001"
    ```

```
        categories: ["S3"]
        supported_frameworks: ["terraform"]
        language: "rego"
        scope: "resource"
        block_type: "aws_s3_bucket"
        inspect_block: |
          package main

          default allow = false

          allow {
            input.acl == "private"
          }
```

2. Run `checkov -f main.tf` to scan your Terraform script with Checkov.

This custom policy checks that all S3 buckets are private. If a bucket is not private, Checkov will report a violation of this policy.

Advanced features of Checkov

Checkov is not just a static code analyzer; it also offers several advanced features that can further enhance your IaC security posture. Let's explore some of these features:

- **Dependency graph**: Checkov builds a graph that represents the relationships between resources in your IaC scripts. This allows Checkov to evaluate not just individual resources, but also the interactions between them. For example, it can detect whether a security group rule allows traffic to a database from an insecure source.

- **Baseline scanning**: Checkov allows you to create a baseline of your current IaC security posture. You can then compare future scans against this baseline to track your progress and ensure that your security posture is improving over time.

- **Policy suppression**: If there are certain policies that you don't want to apply to your IaC scripts, you can suppress them using Checkov. This allows you to customize the security checks that Checkov performs to suit your specific needs.

- **Integration with the Bridgecrew platform**: Checkov integrates with the Bridgecrew platform, which provides additional features such as a policy library, automated fixes, and collaboration tools.

Using Checkov's dependency graph

Let's learn how we can use Checkov's dependency graph to detect a security issue:

1. Create a file named `main.tf` with the following content:

```
provider "aws" {
  region = "us-west-2"
}

resource "aws_security_group" "web" {
  name        = "web"
  description = "Allow all inbound traffic"

  ingress {
    from_port   = 0
    to_port     = 0
    protocol    = "-1"
    cidr_blocks = ["0.0.0.0/0"]
  }
}

resource "aws_db_instance" "default" {
  allocated_storage    = 10
  engine               = "mysql"
  engine_version       = "5.7"
  instance_class       = "db.t2.micro"
  name                 = "mydb"
  username             = "foo"
  password             = "foobarbaz"
  parameter_group_name = "default.mysql5.7"
  publicly_accessible  = true
  vpc_security_group_ids = [aws_security_group.web.id]
}
```

2. Run `checkov -f main.tf` to scan your Terraform script with Checkov.

Checkov will report that the security group allows all inbound traffic and that it is associated with a publicly accessible database. This is a security risk as it allows anyone to access your database.

To fix this vulnerability, you should restrict the inbound traffic to only the necessary ports and IP addresses and make the database not publicly accessible.

Integrating Checkov into your CI/CD pipeline

One of the key benefits of Checkov is its ability to be integrated into your CI/CD pipeline. This allows you to automate the process of checking your IaC scripts for security and compliance issues, ensuring that these checks are performed consistently and that no issues are overlooked. By integrating Checkov into your CI/CD pipeline, you can do the following:

- **Catch security issues early**: By scanning your IaC scripts before they're deployed, you can catch and fix security issues early in the development process. This can save time and effort compared to fixing issues after deployment.

- **Prevent insecure configurations from being deployed**: If Checkov detects a security issue in your IaC script, it can fail the build and prevent the insecure configuration from being deployed.

- **Automate security checks**: By automating your security checks, you can ensure that they're performed consistently and that no issues are overlooked.

Let's learn how we can integrate Checkov into a CI/CD pipeline using GitHub Actions:

1. Create a new GitHub repository and push your Terraform scripts to it.

2. In your repository, create a file named `.github/workflows/checkov.yml` with the following content:

```
name: Checkov

on:
  push:
    branches:
      - main

jobs:
  checkov:
    runs-on: ubuntu-latest

    steps:
    - name: Check out code
      uses: actions/checkout@v2

    - name: Set up Python
      uses: actions/setup-python@v2
      with:
        python-version: '3.x'
```

```
    - name: Install Checkov
      run: pip install checkov

    - name: Run Checkov
      run: checkov -d .
```

3. Push the `.github/workflows/checkov.yml` file to your repository.

Now, whenever you push to the `main` branch of your repository, GitHub Actions will run Checkov against your Terraform scripts.

I strongly encourage you to install the tool and try out multiple scans for IaC scripts as it can be readily integrated into your production environment so that you can hit the ground running with your IaC scanning.

Policy as Code

PaC is a practice in DevOps where policies that govern IT systems and infrastructure are defined and managed as code. This approach allows for automated enforcement and compliance monitoring of these policies, ensuring that all infrastructure deployments are in line with organizational standards and regulatory requirements.

PaC is a key component of a DevSecOps strategy as it allows for security and compliance checks to be automated and integrated into the CI/CD pipeline. This ensures that security is not an afterthought, but an integral part of the infrastructure development process.

Why Policy as Code?

The importance of PaC cannot be overstated. Here's why:

*   **Consistency**: By defining policies as code, you can ensure that they are applied consistently across all of your infrastructure. This eliminates the risk of human error and ensures that all deployments are in line with your policies.

*   **Automation**: PaC allows you to automate policy enforcement and compliance checks. This not only saves time and effort but also ensures that no issues are overlooked.

*   **Version control**: Just like any other code, policy code can be version controlled. This allows you to track changes over time, roll back to previous versions if necessary, and collaborate with your team on policy development.

*   **Documentation**: Policy code serves as a form of documentation, clearly outlining the rules that govern your infrastructure. This can be particularly useful for new team members or for audit purposes.

Implementing Policy as Code with OPA

OPA is an open source, general-purpose policy engine that enables unified, context-aware policy enforcement across the entire stack. We covered this tool in great depth in *Chapter 3* (I strongly encourage you to read that chapter beforehand to gain a deeper understanding of the tool before continuing). Let's see how we can use OPA to implement PaC:

1. Install OPA on your local machine.

2. Create a file named `policy.rego` with the following content:

   ```
   package main

   default allow = false

   allow {
     input.kind == "Deployment"
     input.spec.replicas <= 5
   }
   ```

3. Run the following command to evaluate your policy:

   ```
   $ opa eval --data policy.rego --input input.json "data.main.
   allow"
   ```

This policy checks that Kubernetes deployments have no more than five replicas. If a deployment has more than five replicas, OPA will report a violation of this policy.

Policy as Code in the broader DevSecOps strategy

In the context of DevSecOps, PaC plays a crucial role in integrating security into the development life cycle. It allows security and compliance checks to be automated, ensuring that these checks are performed consistently and that no issues are overlooked. This is particularly important in a cloud-native environment, where infrastructure is often defined and managed as code.

Integrating Policy as Code into the CI/CD pipeline

One of the key benefits of PaC is its ability to be integrated into the CI/CD pipeline. This allows for automated enforcement and compliance monitoring of policies, ensuring that all infrastructure deployments are in line with organizational standards and regulatory requirements.

By integrating PaC into your CI/CD pipeline, you can do the following:

- **Catch security issues early**: By evaluating policies before infrastructure is deployed, you can catch and fix security issues early in the development process. This can save time and effort compared to fixing issues after deployment.

- **Prevent insecure configurations from being deployed**: If a policy violation is detected in your IaC scripts, the build can be failed, and the insecure configuration can be prevented from being deployed.

- **Automate security checks**: By automating your security checks, you can ensure that they're performed consistently and that no issues are overlooked.

Let's see how we can integrate PaC into a CI/CD pipeline using GitHub Actions:

1. Create a new GitHub repository and push your IaC scripts to it.

2. In your repository, create a file named `.github/workflows/policy_check.yml` with the following content:

```yaml
name: Policy Check

on:
  push:
    branches:
      - main

jobs:
  policy_check:
    runs-on: ubuntu-latest

    steps:
    - name: Check out code
      uses: actions/checkout@v2

    - name: Set up OPA
      run: |
        curl -L -o opa https://openpolicyagent.org/downloads/latest/opa_linux_amd64
        chmod 755 ./opa

    - name: Run Policy Check
      run: ./opa eval --data policy.rego --input main.tf "data.main.allow"
```

3. Push the `.github/workflows/policy_check.yml` file to your repository.

Now, whenever you push to the `main` branch of your repository, GitHub Actions will run your policy against your IaC scripts.

Policy as Code – a pillar of DevSecOps

PaC is not just a tool or a practice; it's a fundamental pillar of the DevSecOps philosophy. It embodies the idea that security should be integrated into every stage of the development process, rather than being an afterthought or a separate process.

In a DevSecOps environment, everyone is responsible for security. Developers, operations staff, and security teams all work together to ensure that the infrastructure is secure and compliant. PaC facilitates this collaboration by providing a clear, codified set of rules that everyone can understand and follow.

Policy as Code and Infrastructure as Code – two sides of the same coin

PaC and IaC are closely related concepts. While IaC allows you to define and manage your infrastructure as code, PaC allows you to define and enforce your security and compliance policies as code.

By using IaC and PaC together, you can ensure that your infrastructure is not only defined in a consistent, repeatable manner but also that it adheres to all of your security and compliance requirements. Now that we are aware we can use Terraform as an automation tool for hardening security configurations in the cloud, let's look into another important use case of Terraform – that is, defining custom policies in line with your organization's compliance requirements.

Using Policy as Code with Terraform

Let's see how we can use PaC with Terraform:

1. Install Terraform and OPA on your local machine.
2. Create a file named `main.tf` that contains your Terraform configuration.
3. Create a file named `policy.rego` that contains your OPA policy.
4. Run `terraform plan -out=tfplan` to generate a Terraform plan.
5. Run `terraform show -json tfplan > tfplan.json` to convert the Terraform plan into JSON.
6. Run `opa eval --data policy.rego --input tfplan.json "data.main.allow"` to evaluate your policy against the Terraform plan.

This will check your Terraform plan against your policy, ensuring that the planned infrastructure changes are in line with your security and compliance requirements.

Let's look into each component that builds up to a complete cloud-native infrastructure and understand how DevSecOps can be integrated into each unit of your cloud infrastructure. Remember, DevSecOps isn't all about automation and improving your CI/CD pipeline but is also about a process and mindset change from building software using agile methodologies to integrating security into each.

Container security

In the cloud-native world, containers have become a fundamental unit of deployment. Containers package an application and its dependencies into a single, self-contained unit that can run anywhere, making them ideal for deploying microservices. However, this also introduces new security challenges.

Containers share the host system's kernel, which means that a vulnerability in the kernel could potentially compromise all containers running on that host. Therefore, it's crucial to keep the host system secure and up to date. Using minimal base images for your containers can reduce the attack surface and limit the potential impact of vulnerabilities.

Container images should be scanned for vulnerabilities before deployment. Tools such as Clair and Anchore can be integrated into your CI/CD pipeline to automatically scan images as part of the build process. Runtime security is also important – tools such as Falco can detect anomalous behavior in running containers, alerting you to potential security incidents.

Secrets management

Secrets such as API keys, passwords, and certificates need to be securely managed. They should never be hard-coded into your application code or container images. Instead, they should be securely stored and accessed only when needed.

It's recommended that you use tools such as HashiCorp's Vault instead of Kubernetes Secrets to provide secure storage and access control for secrets. They allow secrets to be encrypted at rest and in transit and provide fine-grained access control to ensure that only authorized applications and users can access them.

> **Important note**
>
> By default, Kubernetes secrets are not encrypted – they are only stored as Base64-encoded versions of the plaintext data and it is always recommended to create an `encryptionConfiguration` file after creating your Kubernetes cluster to enforce encryption at rest for secrets and other sensitive resources within your cluster, along with configuring RBAC to restrict access to secrets to users and other containers.

Network policies

In a cloud-native environment, controlling the traffic flow between services is crucial for security. By default, all the pods in a Kubernetes cluster can communicate with all other pods, regardless of where they are running. This means that if an attacker manages to compromise one pod, they could potentially use it as a launching pad to attack other pods.

Network policies allow you to control which pods can communicate with each other, reducing the potential attack surface. They work at the level of the pod, the basic building block of Kubernetes, and use labels to select pods and define rules that specify what traffic is allowed to the selected pods.

For example, you might have a policy that allows traffic from any pod in the same namespace but blocks traffic from other namespaces. Or you might have a policy that allows traffic only from certain IP addresses. The possibilities are vast and can be tailored to your specific needs.

Defining network policies is a proactive measure that secures your Kubernetes environment. It's a way of applying the principle of least privilege to network traffic: each pod should be allowed to communicate only with the pods it needs to, and no more.

In Kubernetes, network policies are defined as YAML files that specify which pods can communicate with each other. By default, all pods can communicate with all other pods, so it's important to define network policies that restrict communication to only what's necessary for your application to function.

Security in serverless architectures

Serverless architectures, where you only manage your application code and the cloud provider manages the underlying infrastructure, are becoming increasingly popular. However, they also introduce unique security considerations.

In a serverless architecture, each function runs in its own isolated environment, which can limit the potential impact of vulnerabilities. However, it's still important to follow best practices such as using the principle of least privilege for function permissions, validating input data to prevent injection attacks, and monitoring function activity for potential security incidents.

One of the key security considerations in a serverless architecture is the permissions granted to the serverless functions. Each function should be granted only the permissions it needs to perform its task and no more. This can be challenging as it requires a detailed understanding of what each function does and what resources it needs to access.

Input validation is another important consideration. Serverless functions often interact with a variety of inputs, such as HTTP requests, event data, and database records. These inputs can be a vector for attacks if they are not validated properly.

Monitoring function activity can help you detect and respond to security incidents. Many cloud providers provide monitoring and logging services that can be used to track function invocations and detect anomalous behavior.

In a serverless architecture, each function runs in its own isolated environment, which can limit the potential impact of vulnerabilities. However, it's still important to follow best practices such as using the principle of least privilege for function permissions, validating input data to prevent injection attacks, and monitoring function activity for potential security incidents.

Security observability

Monitoring and logging are crucial for detecting and responding to security incidents. They provide visibility into your application's behavior and allow you to detect anomalous or malicious activity.

Tools such as Prometheus for monitoring, Fluentd for logging, and OpenTelemetry for tracing can provide comprehensive observability for your cloud-native applications. These tools can be integrated into your application code and infrastructure, providing real-time insights into your application's security posture.

Prometheus, for example, is an open source monitoring and alerting toolkit that can collect metrics from your applications and infrastructure. These metrics can be used to create dashboards, set up alerts, and troubleshoot issues.

Fluentd is an open source data collector that can collect logs from various sources, transform them, and send them to a variety of destinations. It can be used to centralize your logs, making it easier to search and analyze them.

OpenTelemetry is a set of APIs, libraries, agents, and instrumentation that can be used to collect and manage telemetry data (metrics, logs, and traces) from your applications. It provides a unified way to collect and export telemetry data, making it easier to observe your applications.

Compliance auditing

Maintaining compliance with various standards and regulations is a key requirement for many organizations. In a cloud-native environment, where infrastructure is often defined as code, compliance auditing can be automated.

Tools such as kube-bench for the CIS Kubernetes Benchmark and OPA for policy-based control can be used to automatically check your infrastructure for compliance with various standards. These tools can be integrated into your CI/CD pipeline, providing continuous compliance assurance.

kube-bench, for example, is a Go application that checks whether Kubernetes is deployed according to the security best practices defined in the CIS Kubernetes Benchmark.

OPA, on the other hand, is a general-purpose policy engine that can be used to enforce policies across your stack. With OPA, you can write policies as code, test them, and integrate them into your CI/CD pipeline. This allows you to catch compliance issues early before they make it into production.

Threat modeling and risk assessment

Understanding potential threats and assessing risks is a crucial part of any security strategy. Threat modeling involves identifying potential threats to your application and infrastructure, whereas risk assessment involves evaluating the potential impact and likelihood of these threats.

In a cloud-native environment, threat modeling and risk assessment should be part of your development process. They can help you identify potential security issues early when they are easier and less costly to fix.

Threat modeling typically involves creating a diagram of your application and infrastructure, identifying potential threats, and determining what controls are in place to mitigate those threats. This can be a manual process, but there are also tools available that can automate parts of it.

Risk assessment, on the other hand, involves evaluating the potential impact of each identified threat and the likelihood of it occurring. This can help you prioritize your security efforts and focus on the most significant risks.

Incident response

Despite your best efforts, security incidents can still occur. Having a plan for responding to these incidents is crucial. This plan should include steps for identifying, containing, and resolving the incident, as well as communicating with stakeholders and learning from the incident to prevent future occurrences.

In a cloud-native environment, incident response can be more challenging due to the ephemeral and distributed nature of the infrastructure. However, tools and practices such as centralized logging, automated alerts, and chaos engineering can help you prepare for and respond to incidents effectively.

Centralized logging can help you quickly identify and investigate incidents. By collecting logs from all parts of your infrastructure in one place, you can easily search for and analyze them to determine what happened.

Automated alerts can help you detect incidents quickly. By setting up alerts based on metrics or log patterns, you can be notified of potential incidents as soon as they occur.

Chaos engineering, the practice of intentionally introducing failures into your system to test its resilience, can help you prepare for incidents. By regularly testing your system's ability to handle failures, you can identify and fix issues before they cause a real incident.

Security training and culture

DevSecOps is as much about people and culture as it is about tools and technology. Building a culture of security within your organization is crucial for the success of your DevSecOps initiative.

This involves providing security training for all members of your organization, not just the security team. Everyone should understand the importance of security and their role in maintaining it. Regularly reviewing and updating your security policies and practices, and encouraging open communication about security issues, can also help build a strong security culture.

In conclusion, DevSecOps is a comprehensive approach to security that involves integrating security practices into every stage of the development process. By using tools and practices such as PaC, IaC, container security, secrets management, network policies, security observability, compliance auditing, threat modeling and risk assessment, incident response, and security training and culture, you can build secure and compliant cloud-native applications.

Continuous learning and improvement – the DevSecOps mindset

DevSecOps is a mindset of continuous learning and improvement. As new technologies emerge and threats evolve, so too must our DevSecOps practices. This requires a commitment to ongoing education and a willingness to adapt and change.

One way to foster continuous learning is through regular training and professional development opportunities. This could include attending conferences, participating in webinars, or taking online courses. There are many resources available to help you stay up to date with the latest developments in DevSecOps and cloud-native security.

Another way to promote continuous improvement is through regular reviews and audits of your DevSecOps practices. This could involve conducting internal audits, seeking external audits, or using automated tools to continuously check for security misconfigurations and vulnerabilities. The goal is to identify areas of weakness and opportunities for improvement.

The role of automation in DevSecOps

Automation plays a crucial role in DevSecOps, particularly in a cloud-native environment. It not only increases efficiency and consistency but also reduces the risk of human error, which is often a major cause of security incidents.

In the context of DevSecOps, automation can be applied to various stages of the software development life cycle. For instance, automated testing tools can be used to identify security vulnerabilities in the code during the development phase. Similarly, automated deployment tools can ensure that the application is deployed securely and consistently across different environments.

Moreover, automation can also be used to enforce security policies. For example, PaC tools can automatically enforce security policies in the CI/CD pipeline, preventing insecure code or configurations from being deployed.

The importance of collaboration in DevSecOps

DevSecOps is not just about tools and technologies; it's also about people and processes. It requires close collaboration between development, security, and operations teams. This collaboration is often referred to as a "security culture."

In a security culture, everyone in the organization understands the importance of security and takes responsibility for it. Security is not seen as someone else's job but as a shared responsibility. This culture is often characterized by open communication, shared learning, and mutual respect.

Building a security culture requires leadership support, ongoing education, and clear communication. It also requires tools and processes that facilitate collaboration and make it easy for everyone to do their part in maintaining security.

The power of open source in DevSecOps

Open source software plays a significant role in DevSecOps. Many of the tools used in DevSecOps, such as Kubernetes, Terraform, and Checkov, are open source. These tools are not only free to use but also benefit from the collective wisdom and effort of a global community of developers and security professionals.

Using open source tools can provide several benefits. For one, it can reduce costs as there are no licensing fees. It can also increase flexibility as you can modify the tools to suit your specific needs. Moreover, open source tools are often more transparent and trustworthy as their source code is publicly available for review.

However, using open source tools also comes with responsibilities. You need to ensure that the tools are secure and up to date. You also need to comply with their licensing terms, which may require you to contribute back to the community.

Future trends – the evolution of DevSecOps

As we look to the future, several trends are likely to shape the evolution of DevSecOps. These include the increasing adoption of machine learning and artificial intelligence for security, the rise of zero-trust architectures, and the growing importance of privacy and data protection.

Machine learning and artificial intelligence can help automate and enhance many aspects of DevSecOps, from threat detection to incident response. However, they also introduce new challenges and require careful consideration in terms of ethical and privacy implications.

Zero-trust architectures, which assume that any part of the system could be compromised and therefore require continuous verification, are becoming increasingly popular as a way to enhance security in cloud-native environments.

Privacy and data protection are also becoming increasingly important, driven by both regulatory requirements and consumer demand. This requires a holistic approach to security that considers not only the technical aspects but also the human and organizational aspects.

Summary

In conclusion, DevSecOps is a dynamic and evolving field that requires continuous learning and improvement. By embracing this mindset and staying engaged with the community, you can help secure your cloud-native applications and contribute to the advancement of DevSecOps practices.

In this chapter, we explored the various facets of DevSecOps practices in a cloud-native environment. We discussed the importance of IaC and Policy as Code, and how they contribute to the security of cloud-native applications and infrastructure. We also delved into the various open source tools hosted by the CNCF that are used to secure the pipeline and code development.

We also discussed the importance of container security, secrets management, network policies, security in serverless architectures, security observability, compliance auditing, threat modeling and risk assessment, incident response, and security training and culture.

By understanding and implementing these practices, you will be well on your way to securing your cloud-native applications and infrastructure. Remember, security is not a one-time event, but a continuous process that requires ongoing attention and improvement.

I hope that this chapter has provided you with a solid foundation in DevSecOps practices for cloud-native applications. I encourage you to continue exploring these topics and applying what you've learned in your own projects. In the next chapter, we will learn about certain laws, regulations, doctrines, and policies established by government entities that you should be following and how to use automated tools such as Terraform to ensure that your application and your organization as a whole are compliant with those policies.

Quiz

Answers the following questions to test your knowledge of this chapter:

- How does integrating DevSecOps practices into the CI/CD pipeline enhance the security of cloud-native applications? Can you think of a scenario where this integration could have prevented a security incident?

- In the context of IaC and Policy as Code, how do tools such as Terraform and Checkov contribute to the security of cloud-native applications? Can you envision a situation where the use of these tools could lead to a security vulnerability if they're not managed properly?

- How does a security culture within an organization contribute to the success of DevSecOps practices? Can you provide an example of how a strong security culture could influence the outcome of a security incident?

- Considering the role of open source tools in DevSecOps, what are the potential benefits and challenges of using these tools in a cloud-native environment? Can you think of a situation where the use of an open source tool significantly impacted the security posture of an application?

- Reflecting on the future trends in DevSecOps, such as the adoption of machine learning and artificial intelligence for security, the rise of zero-trust architectures, and the growing importance of privacy and data protection, how do you think these trends will shape the evolution of DevSecOps? Can you predict a potential security challenge that might arise with the adoption of these trends?

Further readings

To learn more about the topics that were covered in this chapter, take a look at the following resources:

- `https://landscape.cncf.io/`

- `https://www.terraform.io/docs/index.html`

- `https://www.checkov.io/1.Welcome/What%20is%20Checkov.html`

- `https://kubernetes.io/docs/concepts/services-networking/network-policies/`

- `https://www.hashicorp.com/resources/what-is-mutable-vs-immutable-infrastructure`

- `https://www.cncf.io/blog/2019/03/18/demystifying-containers-part-iv-container-security/`

- `https://www.cncf.io/blog/2019/04/18/demystifying-containers-part-vi-container-security-orchestration-and-deployment/`

- `https://kubernetes.io/docs/concepts/configuration/secret/#:~:text=Kubernetes%20Secrets%20are%2C%20by%20default,anyone%20with%20access%20to%20etcd`

Part 3: Legal, Compliance, and Vendor Management

In this part, you will learn about the laws, regulations, and standards that govern your work in cloud-native security. You will also gain insights into cloud vendor management and security certifications. By the end of *Part 3*, you will understand the various risks associated with cloud vendors and how to assess a vendor's security posture effectively.

This part has the following chapters:

9

Legal and Compliance

In the world of **cloud-native software security**, understanding the legal and compliance aspects is as crucial as mastering the technical skills. This chapter aims to bridge the gap between these two seemingly disparate areas, providing you, the security engineer, with a comprehensive understanding of the laws, regulations, and standards that govern your work. In the previous chapter, you learned about different techniques for automating security and compliance policies using DevSecOps tools such as Terraform and other incident response tools. Following the same train of thought, in this chapter, we will learn about even more policies and compliance standards set out by government bodies across different countries that tech companies are expected to follow.

By the end of this chapter, you will not only have gained knowledge about the key US privacy and security laws but have also learned how to analyze these laws from a security engineer's perspective. We will delve into real-world case studies that illustrate the practical implications of these laws, offering you a unique perspective on how they impact your day-to-day operations.

Moreover, we will explore the audit processes and methodologies that are integral to **cloud-native adoption**, helping you understand how to integrate legal and compliance considerations into your **DevSecOps** practices. This knowledge will empower you to build more secure, compliant, and resilient systems, thereby enhancing your value proposition as a security professional. This chapter promises to be a unique journey, blending the worlds of law and cloud-native software security in a way that is both informative and engaging.

In this chapter, we're going to cover the following main topics:

- Overview of this chapter
- Comprehending privacy in the cloud
- Audit processes, methodologies, and cloud-native adoption
- Laws, regulations, and standards

Overview

In the rapidly evolving landscape of cloud-native software security, the importance of understanding the legal and compliance aspects cannot be overstated. As a security engineer, you're not just tasked with building secure systems but also ensuring that these systems comply with a complex web of laws, regulations, and standards. This chapter is designed to demystify these legal and compliance aspects, providing you with a clear, concise, and comprehensive understanding of the topic.

We will begin with *Comprehending privacy in the cloud*, where we'll explore the key US privacy laws, such as the **California Consumer Privacy Act** (**CCPA**), and their implications for cloud-native software security. We'll break down these laws into simple, understandable terms and illustrate their practical implications through real-world case studies.

Next, in *Audit processes, methodologies, and cloud-native adoption*, we'll delve into the audit processes and methodologies that are crucial to ensure legal and compliance considerations are present in cloud-native adoption. Through practical examples, we'll show you how these processes work and why they're important.

In the *Laws, regulations, and standards* section, we'll take a closer look at the key US security laws, such as the **Computer Fraud and Abuse Act** (**CFAA**) and the **Federal Trade Commission Act** (**FTCA**), as well as important compliance standards such as **Service Organization Control 2** (**SOC 2**) and **the Payment Card Industry Data Security Standard** (**PCI DSS**). We'll explain these laws and standards in a way that's easy to understand and relate them to your work as a security engineer.

Finally, in the *Laws, regulations and standards* section, we'll do a deeper dive into these laws, helping you understand their nuances and how they impact your work.

Comprehending privacy in the cloud

In the digital age, privacy has become a cornerstone of consumer rights and a critical consideration for businesses, especially those operating in the cloud. As a security engineer, understanding privacy laws and their implications for cloud-native software security is crucial. This section will explore key US privacy laws and their implications for cloud-native software security.

Understanding the language of privacy is the first step toward comprehending the complex landscape of privacy laws and regulations. Here, we'll delve deeper into some of the key terms related to privacy:

- **Personal data**: Personal data, also known as **personally identifiable information** (**PII**), refers to any information that can be used to identify an individual. This could include **direct identifiers** such as names, social security numbers, and physical addresses, as well as **indirect identifiers** such as IP addresses, location data, and unique device identifiers. In a **cloud-native environment**, personal data can be stored in various formats and locations, making protecting it a significant challenge.

- **Data controller**: The data controller is the entity that determines the purposes (the *why*) and means (the *how*) of processing personal data. In a cloud-native context, the data controller could be your organization, which is making decisions about what personal data to collect from users, how to use it, and how long to retain it. The data controller is responsible for ensuring that data processing activities comply with applicable privacy laws.

- **Data processor**: The data processor is the entity that processes personal data on behalf of the controller. Processing can include a range of activities, such as collection, recording, organization, structuring, storage, adaptation, retrieval, consultation, use, disclosure, dissemination, alignment, combination, restriction, erasure, or destruction. In a cloud-native environment, data processors could include cloud service providers (CSPs) or other third-party services that handle personal data on your behalf.

- **Data subject**: The data subject is the individual whose personal data is being processed. Data subjects have certain rights under privacy laws, such as the right to access their data, correct inaccuracies, object to processing, and have their data erased under certain circumstances. As a security engineer, you need to ensure that your systems can support these data subject rights.

- **Data breach**: A data breach is a security incident where unauthorized individuals gain access to personal data. This could occur due to various reasons, such as a cyberattack, a system vulnerability, or human error. In a cloud-native environment, the distributed nature of applications and data can increase the risk of data breaches. However, it also provides opportunities for advanced security measures, such as **encryption**, **automated threat detection**, and **incident response**.

By understanding these key terms, you'll be better equipped to navigate the world of privacy and ensure that your cloud-native applications are designed and operated in a manner that respects and protects personal data.

The importance of privacy in the cloud-native landscape

In cloud-native environments, data often moves across different services and infrastructures, increasing the risk of data breaches. As a security engineer, it's your responsibility to ensure that personal data is protected at all stages of this journey. This not only involves implementing robust security measures but also ensuring compliance with privacy laws.

The distributed nature of cloud-native applications, the vast amount of data they handle, and the speed at which they operate make privacy a crucial aspect of their design and operation.

It becomes prudent to start thinking about the privacy constraints for the application that you're trying to develop or protect as soon as you start the application architecture, as it is not only a beneficial step for the consumers and organizations in adhering to the privacy law – it also brings many other benefits of following a robust privacy program. A few of them are as follows:

- **Trust and reputation**: Privacy is a fundamental aspect of building trust with customers. When customers entrust their data to a company, they expect it to be handled with care and respect. Any breach of this trust, such as a data leak or misuse of data, can severely damage a company's reputation and customer relationships. In a cloud-native environment, where data often moves across different services and infrastructures, maintaining privacy is key to preserving this trust.

- **Regulatory compliance**: As we've discussed, there are numerous laws and regulations, such as the CCPA, the Health Insurance Portability and Accountability Act (HIPAA), and the General Data Protection Regulation (GDPR) (this will be explained in more detail in the *Laws, regulations, and standards* section), that mandate certain standards of privacy. Non-compliance with these regulations can result in hefty fines and legal repercussions. Therefore, understanding and adhering to these privacy laws is not just a matter of ethical data handling but also a legal requirement.

- **Competitive advantage**: In today's data-driven world, businesses that can demonstrate robust privacy practices have a competitive edge. Consumers are becoming more aware of their privacy rights and are more likely to choose companies that respect and protect these rights. Therefore, integrating privacy into the core of your cloud-native software security practices can differentiate your business in the market.

- **Mitigating risks**: Privacy breaches can lead to significant financial and reputational damage. In a cloud-native environment, the risk is amplified due to the potential scale of a breach. By prioritizing privacy, you can mitigate these risks, protect your customers, and safeguard your business.

- **Ethical responsibility**: Beyond the legal and business reasons, there's an ethical imperative to protect privacy. As a security engineer, you have a responsibility to ensure that the technologies you build and deploy respect users' privacy rights. This ethical responsibility is becoming increasingly recognized in the tech industry and is an integral part of responsible cloud-native software security.

In conclusion, privacy is not just a legal requirement or a nice-to-have feature in cloud-native software security. It's a fundamental aspect of how businesses operate and how they build and maintain trust with their customers. As a security engineer, understanding and prioritizing privacy is a crucial part of your role.

Now that we understand the importance of privacy for the organizations in their applications, let's look into one such very important US privacy law that any company operating in California or with a substantial user base of their app in California has to adhere to.

The CCPA and its implications for cloud-native

The CCPA is a landmark privacy law that came into effect in California in 2020. It was designed to enhance privacy rights and consumer protection for residents of California, United States. The law has wide-ranging implications for companies that do business in California or handle the personal data of California residents.

Key principles of the CCPA

The CCPA is built on the following key principles, which reflect a shift toward greater consumer control over personal data:

- **Transparency**: The CCPA requires businesses to be transparent about their data collection practices. They must inform consumers, at or before the point of data collection, about the categories of personal information to be collected and the purposes for which these categories will be used. In a cloud-native environment, this could mean clearly stating in your privacy policy what data your application collects, why it collects it, and how it uses and shares this data.

- **Control**: The CCPA empowers individuals with greater authority over their private data. It allows them to ask companies to reveal both the types and specific instances of personal data that the company has gathered about them. Furthermore, it grants them the ability to demand the removal of their data and to choose not to have their personal information sold. For cloud-native applications, this could involve building features that allow users to view, delete, and download their data, and to opt out of data sharing.

- **Accountability**: The CCPA holds businesses accountable for protecting personal information. As per the CCPA regulatory criteria laid out, businesses are required to implement reasonable security procedures and practices to protect personal information from unauthorized access, destruction, use, modification, or disclosure. In the event of certain data breaches, businesses can be held liable. For cloud-native software security, this underscores the importance of implementing robust security measures, such as encryption, access controls, and security monitoring, and of having an incident response plan in place.

Implications for cloud-native software security

The CCPA's principles of transparency, control, and accountability have several implications for cloud-native software security. They are as follows:

- **Data inventory and mapping**: One of the first steps tech organizations have had to take in response to the CCPA is to conduct a thorough data mapping and inventory exercise. This involves identifying what personal information is collected, where it is stored, how it is used, who it is shared with, and how long it is retained. This is a significant task in a cloud-native environment, where data can be distributed across various services and storage locations. Privacy engineers have had to work closely with their teams to develop or adapt data mapping tools and processes to maintain an accurate and up-to-date data inventory.

- **Privacy by design**: The CCPA has also reinforced the importance of the **privacy-by-design approach**, which involves integrating privacy considerations into every stage of the product development process. This means that privacy engineers are now more involved in the early stages of product development, working with product managers, developers, and UX designers to ensure that privacy features are built into products from the ground up. This could involve designing user-friendly privacy notices, developing mechanisms for users to exercise their CCPA rights, or implementing technical measures to protect personal data.

- **Data access and deletion requests**: You need to be able to fulfill data access and deletion requests from data subjects. This requires having systems in place to locate and remove personal data across your cloud-native environment.

- **Vendor management**: The CCPA has also necessitated a closer examination of relationships with third-party vendors. Tech organizations have had to review their contracts with vendors to ensure that they include the necessary provisions for CCPA compliance, such as restrictions on the sale of personal information. Privacy engineers have played a crucial role in this process, working with legal and procurement teams to assess vendor compliance and negotiate contract terms.

- **Data protection**: The CCPA's requirement for reasonable security procedures and practices has underscored the importance of robust data protection measures. Privacy engineers have been at the forefront of implementing and enhancing these measures, which could include encryption, access controls, network security measures, and security monitoring tools. They have also been involved in developing incident response plans to ensure a swift and effective response to any data breaches.

- **Training and awareness**: Finally, the CCPA has highlighted the need for privacy awareness and training across organizations. Privacy engineers are often tasked with developing and delivering this training, ensuring that all employees understand the importance of privacy and their responsibilities under the CCPA.

In conclusion, the CCPA is a significant privacy law that has reshaped the privacy landscape in the US and has important implications for cloud-native software security. As a security engineer, understanding the CCPA and integrating its principles into your work is crucial to ensure compliance and build trust with users.

Other significant US privacy laws and their implications for cloud-native

While the CCPA has been a significant milestone in the US privacy landscape, it's not the only privacy law that impacts cloud-native software security. Several other states have enacted or are considering similar laws, and there are also sector-specific privacy laws that you need to be aware of.

Nevada's online privacy Law (Senate Bill 220)

Nevada's online privacy law, also known as **Senate Bill 220**, is another state-level privacy law that has implications for cloud-native software security. Like the CCPA, it gives consumers the right to opt out of the sale of their personal information. However, it differs from the CCPA in several ways. For example, it has a narrower definition of *sale*, and it doesn't provide a general right to access or delete personal information.

For cloud-native security, the key implication of Nevada's law is the need to provide a designated request address (for example, a toll-free number or email address) where consumers can submit opt-out requests. You also need to respond to these requests within 60 days. This requires a robust system for managing and responding to consumer requests, which can be particularly challenging in a cloud-native environment due to the distributed nature of data and services.

Virginia's Consumer Data Protection Act (CDPA)

Virginia's **Consumer Data Protection Act (CDPA)** is another significant privacy law. It shares many similarities with the CCPA and the GDPR, but it also has some unique provisions. For example, it introduces the concept of *data protection assessments*, which are risk assessments that certain businesses need to conduct when processing personal data in specified high-risk situations.

For cloud-native security, the CDPA emphasizes the importance of **data minimization, purpose limitation**, and **security**. It also requires businesses to establish, implement, and maintain reasonable administrative, technical, and physical data security practices to protect the confidentiality, integrity, and accessibility of personal data. This reinforces the need for robust security measures in a cloud-native environment, such as **encryption, access controls**, and **security monitoring**.

Sector-specific privacy laws

In addition to these state-level laws, there are also several sector-specific privacy laws in the US, such as the HIPAA for healthcare information, the **Gramm-Leach-Bliley Act (GLBA)** for financial information, and the **Children's Online Privacy Protection Act (COPPA)** for children's information.

These laws have specific requirements for protecting certain types of personal information, and they can have significant implications for cloud-native software security. For example, if you're handling protected health information in a cloud-native environment, you need to ensure that you're complying with **HIPAA's Security Rule**, which requires specific administrative, physical, and technical safeguards. This could involve implementing encryption for data at rest and in transit, using secure coding practices, and conducting regular security audits.

Similarly, if you're handling financial information, you need to comply with the **GLBA's Safeguards Rule**, which requires financial institutions to develop a written information security plan that describes how they protect customer information. This could involve identifying and assessing risks to customer information, designing and implementing a safeguarding program, and regularly monitoring and testing the program (you can read more on the GLBA by following the link provided in the *Further reading* section).

Now that we understand the major laws established by US federal agencies, let's try to understand the audit processes that enforce the use of these laws and policies.

Audit processes, methodologies, and cloud-native adoption

Audits play a critical role in ensuring that cloud-native technologies are implemented and used in a manner that complies with various privacy and security laws, regulations, and standards. They also help identify and manage risks, build trust with stakeholders, and drive continuous improvement in privacy and security practices.

In this section, we will delve into the importance of audit processes and methodologies in the context of cloud-native adoption. We will discuss how audits can help ensure compliance, manage risks, build trust, and drive improvement. We will also provide an overview of common audit processes and methodologies, including **internal and external audits**, **self-assessments**, **regulatory audits**, and various **audit methodologies**. We will also explore the tools and technologies that can support audit processes in cloud-native environments.

Whether you are a software engineer, a security engineer, or a privacy professional, understanding audit processes and methodologies is crucial to ensure that your cloud-native applications are secure, privacy-compliant, and trustworthy. So, let's dive in and explore these topics in detail.

Importance of audit processes and methodologies in cloud-native adoption

The adoption of cloud-native technologies brings with it a host of benefits, including scalability, flexibility, and speed. However, it also introduces new complexities and challenges when it comes to ensuring privacy and security. This is where audit processes and methodologies come into play. They serve as critical tools in managing these challenges and ensuring that cloud-native technologies are used responsibly and effectively. Compliance teams usually drive an annual audit process across different teams in an organization for the following reasons.

Ensuring compliance

One of the primary roles of audit processes is to ensure compliance with various privacy and security laws, regulations, and standards. In the US, these may include state-level laws such as the CCPA, sector-specific laws such as the HIPAA, and standards such as the PCI DSS. Internationally, regulations such as the GDPR in the European Union also come into play.

Compliance audits typically involve a systematic review of an organization's policies, procedures, and systems to ensure they meet the requirements of these laws and standards. In a cloud-native environment, this could involve reviewing how personal data is collected, stored, processed, and shared across various services and infrastructures. It could also involve assessing the security measures in place to protect this data, such as encryption, access controls, and incident response plans.

Risk management

Audits also play a crucial role in **risk management**. They can help identify security risks and vulnerabilities in cloud-native environments, enabling organizations to take corrective action before these issues lead to data breaches or other security incidents.

Risk-based audits typically involve a systematic process of identifying and assessing risks, evaluating the effectiveness of controls in mitigating these risks, and recommending improvements. In a cloud-native environment, this could involve identifying risks associated with specific technologies or architectures, such as **container orchestration systems** or **serverless functions**. It could also involve assessing risks associated with third-party services, **APIs**, or open source components used in the environment.

Building trust

Regular audits can also help build trust with customers, stakeholders, and regulators. By demonstrating that the organization takes privacy and security seriously and has robust controls in place, audits can enhance the organization's reputation and credibility.

Trust-building audits often involve not only assessing the organization's compliance with laws and standards but also evaluating its adherence to best practices and ethical guidelines. In a cloud-native environment, this could involve assessing how the organization manages user consent, how it responds to data access or deletion requests, or how it handles data breaches.

Continuous improvement

Finally, audit findings can drive continuous improvement in cloud-native software security practices and processes. By identifying gaps or weaknesses, audits can provide valuable insights and recommendations for enhancing privacy and security.

Continuous improvement audits often involve a cyclical process of **planning, doing, checking, and acting** (the **PDCA cycle**). In a cloud-native environment, this could involve planning improvements based on audit findings, implementing these improvements, checking their effectiveness through follow-up audits, and acting on the results to make further improvements.

Common audit processes and methodologies

In the realm of privacy and security, audits serve as a vital mechanism for ensuring that an organization's practices align with its stated policies, as well as with applicable laws and regulations. There are several types of audits, each with its own focus and methodology. Let's delve into these in detail.

Internal audits

Internal audits are conducted by an organization's audit team. These audits assess the organization's compliance with its policies and procedures, as well as with applicable laws and regulations. In a cloud-native environment, internal audits might involve reviewing the security configurations of cloud services, assessing the handling of personal data across microservices, or evaluating the effectiveness of incident response procedures. Internal audits allow for a thorough, in-depth review of practices and provide an opportunity for the organization to identify and address issues proactively.

External audits

External audits, on the other hand, are typically conducted by independent third parties. These audits provide an objective assessment of the organization's privacy and security practices. In a cloud-native context, an external audit might involve reviewing the organization's cloud security architecture, assessing its compliance with standards such as **ISO 27001** or **SOC 2**, or evaluating its data protection practices. External audits can provide valuable third-party validation of an organization's practices, enhancing its credibility and trustworthiness in the eyes of customers, partners, and regulators.

Self-assessments

Self-assessments enable organizations to proactively evaluate their own privacy and security practices. These assessments can be particularly useful in a rapidly evolving environment such as cloud-native, where new technologies, architectures, and practices are continually being adopted. A self-assessment might involve a review of the organization's cloud-native security strategy, an evaluation of its data privacy practices, or a test of its incident response capabilities. Self-assessments can help the organization identify areas of strength and weakness, and they can inform ongoing improvement efforts.

Regulatory audits

Regulatory audits focus on ensuring compliance with specific laws and regulations, such as the CCPA or HIPAA. These audits are often conducted by regulators or independent auditors acting on their behalf. In a cloud-native environment, a regulatory audit might involve a detailed review of the organization's data handling practices, an assessment of its security controls, or an evaluation of its compliance with specific regulatory requirements. Regulatory audits can help the organization demonstrate its compliance with the law, avoid penalties, and build trust with regulators.

Audit methodologies

Several common **audit methodologies** can be applied in a cloud-native context. The **Risk-Based Audit Approach** focuses on the areas of highest risk to the organization. The **Process-Based Audit Approach** focuses on the organization's processes, looking at how they are designed and how they operate in practice. The **Compliance-Based Audit Approach** focuses on the organization's compliance with specific laws, regulations, or standards. Each of these approaches has its strengths and can be chosen based on the specific objectives of the audit.

Audit tools and technologies

A variety of tools and technologies can support audit processes in cloud-native environments. **Automated audit tools** can help identify security vulnerabilities, monitor compliance with policies, and track changes that have been made to configurations. **Data analytics** can help analyze large volumes of audit data, identify patterns and trends, and generate insights. **Artificial intelligence** and **machine learning** can help automate parts of the audit process, such as **risk assessment** or **anomaly detection**. These tools and technologies can make the audit process more efficient and effective, and they can help organizations keep pace with rapid changes in the cloud-native environment.

Now that we understand the audit process and important US privacy laws, let's do a deep dive into other laws and regulations set up by other government bodies and regulatory departments that set up standards for organizations to adhere to.

Laws, regulations, and standards

These legal and regulatory frameworks provide the rules and guidelines that organizations must follow to ensure the privacy and security of their systems and data. They also establish penalties for non-compliance, which can include fines, sanctions, and even criminal charges:

- **Laws** are enacted by governments and apply to all individuals and entities within their jurisdiction. They establish legal obligations for privacy and security, such as the requirement to protect personal data or to report data breaches.

- **Regulations** are rules issued by governmental agencies under the authority of laws. They provide more detailed requirements for specific sectors or activities. For example, the HIPAA is a law, but the **HIPAA Privacy Rule** and the **HIPAA Security Rule** are regulations issued under the authority of that law.

- **Standards** are guidelines developed by industry or standards-setting organizations. They provide best practices for privacy and security, and compliance with them is often voluntary. However, in some cases, compliance with certain standards can be a legal or regulatory requirement, or it can be used as evidence of due diligence in the event of a legal dispute.

In this section, we will delve into some of the key laws, regulations, and standards that apply to cloud-native software security. We will discuss their key principles, their implications for cloud-native environments, and real-world cases that illustrate these implications. Whether you are a software engineer, a security engineer, or a privacy professional, understanding these legal and regulatory frameworks is crucial for ensuring that your cloud-native applications are secure, compliant, and trustworthy.

The CFAA and its implications for cloud-native software security

The **CFAA** is a federal cybersecurity law in the US that provides a legal framework for prosecuting cybercrimes. The law, enacted in 1986 and subsequently amended several times, is designed to protect computer systems by criminalizing unauthorized access and related activities.

Key principles of the CFAA

The CFAA makes it illegal to intentionally access a computer without authorization or in excess of authorization, and thereby obtain information from any protected computer. The law also prohibits the transmission of harmful code and the trafficking of computer passwords if such conduct affects interstate or foreign commerce.

The CFAA's definition of a *protected computer* is broad, covering any computer used in or affecting interstate or foreign commerce or communication. This includes virtually all computers connected to the internet, making the CFAA applicable to a wide range of activities in cloud-native environments.

For cloud-native environments, the CFAA underscores the importance of implementing robust access controls to prevent unauthorized access to systems and data. It also highlights the need for measures to protect against the transmission of harmful code, such as **malware scanning** and **intrusion detection systems**.

Case study – a CFAA-related incident and its implications for security engineers

One notable case involving the CFAA is *United States v. Nosal, 676 F.3d 854* (the case law is included in the *Further readings* section), in which a former employee of an executive search firm used the login credentials of a current employee to access the firm's database and obtain proprietary information. **The Ninth Circuit Court of Appeals** held that this constituted unauthorized access under the CFAA, even though the current employee had voluntarily provided the login credentials.

This case has significant implications for security engineers in cloud-native environments. It highlights the importance of implementing measures to prevent the sharing of login credentials, such as two-factor authentication and regular password changes. It also underscores the need for clear policies on acceptable use of systems and data, and for measures to monitor and detect unauthorized access.

The FTCA and its implications for cloud-native software security

The **FTCA** is a pivotal piece of legislation in the US that established the **Federal Trades Commission (FTC)** and granted it the authority to enforce antitrust laws and protect consumers. Over time, the FTC has used its authority under the FTCA to address a wide range of privacy and data security issues, making the FTCA a significant law in the realm of cloud-native software security.

Key principles of the FTCA

The FTCA prohibits "*unfair or deceptive acts or practices in or affecting commerce.*" This broad mandate has been interpreted by the FTC to include practices that fail to comply with published privacy policies, that mislead consumers about the security measures in place to protect their personal information, or that fail to provide reasonable security for consumer data.

The FTC has issued guidelines and recommendations on what it considers to be *reasonable* data security practices. While these guidelines are not binding rules, they provide valuable insight into the FTC's expectations and can serve as a benchmark for organizations seeking to avoid enforcement action.

For cloud-native environments, the FTCA underscores the importance of transparency and honesty in dealing with consumers' personal information. It also highlights the need for robust security measures that are commensurate with the sensitivity of the data and the nature of the organization's operations.

Case study – the FTC's prosecution of Uber for lax security policies in AWS key storage

One notable case involving the FTCA and cloud-native software security is the FTC's enforcement action against Uber Technologies, Inc. In 2017, the FTC alleged that Uber had made deceptive claims about its privacy and data security practices. The case stemmed from a data breach in 2014, in which an intruder accessed personal information about Uber drivers stored in an **Amazon Web Services (AWS) S3 data store**.

The FTC's complaint focused on several aspects of Uber's practices. First, the FTC alleged that Uber had failed to live up to its claims of monitoring access to personal information. Second, the FTC alleged that Uber had stored sensitive data in plaintext in AWS, contrary to its claim of using "*the best security technology.*" Finally, the FTC alleged that Uber had failed to provide reasonable security for its databases, noting that Uber had allowed engineers to use a single key that provided full administrative access to all data and that Uber had stored that key in plaintext in its code on **GitHub**, a platform accessible to the public.

The case was settled in 2018, with Uber agreeing to several conditions, including implementing a comprehensive privacy program and undergoing regular, independent audits.

This case has significant implications for security engineers in cloud-native environments. It highlights the importance of accurately representing data security practices to consumers and implementing robust security measures, such as encryption and access controls. It also underscores the risks of storing sensitive data in the cloud without adequate safeguards and the need for careful management of access keys.

Overview of compliance standards and their implications for cloud-native software security

In the world of cloud-native software security, compliance with established standards is not just a matter of best practice – it's often a legal requirement. These standards, set by industry bodies and regulatory agencies, provide a framework for managing and protecting data effectively and securely. Let's explore some of these key standards and their implications for cloud-native environments.

SOC 2

SOC 2 is a type of audit report developed by the **American Institute of Certified Public Accountants (AICPA)**. It's designed to provide assurance that a service organization, such as a CSP, has implemented controls that effectively address **security**, **availability**, **processing integrity**, **confidentiality**, and **privacy**.

SOC 2 reports come in two types – Type I and Type II:

- **SOC 2 Type I**: This type of report focuses on designing controls at a specific point in time. It's essentially a snapshot that describes the systems that a service organization has in place and whether they are designed effectively to meet relevant trust service criteria. A Type I report can be useful for identifying potential issues with a service organization's controls, but it doesn't provide assurance that the controls have operated effectively over time.

- **SOC 2 Type II**: This type of report goes a step further by evaluating the operational effectiveness of controls over a specified period, typically no less than 6 months. It involves more detailed testing and provides a higher level of assurance than a Type I report. A Type II report can provide valuable information for assessing the risks associated with using a service organization's services.

In a cloud-native environment, SOC 2 compliance is particularly important because of the reliance on third-party services. The following are some key considerations:

- **Security**: The system is protected against unauthorized access (both physical and logical). This includes protection against security incidents, system failures, and unauthorized changes to system configurations.

- **Availability**: The system is available for operation and use as committed or agreed. This includes monitoring system performance and availability, incident handling, and disaster recovery.

- **Processing integrity**: System processing is complete, accurate, timely, and authorized. This includes quality assurance and process monitoring.

- **Confidentiality**: Information designated as confidential is protected. This includes data encryption, network and application firewalls, and access controls.

- **Privacy**: Personal information is collected, used, retained, disclosed, and disposed of in conformity with the commitments in the entity's privacy notice and with the criteria outlined in the **Generally Accepted Privacy Principles (GAPP)** issued by the AICPA and the **Canadian Institute of Chartered Accountants (CICA)**. These are the bodies responsible for creating and maintaining the framework for privacy management through the GAPP.

For a security engineer, understanding SOC 2 requirements is crucial. It's not just about passing an audit – it's about ensuring that your systems and data are truly secure and reliable. This involves implementing robust controls, regularly reviewing and updating these controls, and being prepared to demonstrate their effectiveness in an audit.

The PCI DSS

The PCI DSS is a set of security standards developed by the **Payment Card Industry Security Standards Council (PCI SSC)**. It is designed to protect cardholder data and ensure the secure handling of credit card transactions. Compliance with the PCI DSS is essential for any organization that processes, stores, or transmits payment card information.

Here's a comprehensive overview of the PCI DSS and its key components:

- **Scope**: According to the PCI SSC, the PCI DSS applies to all entities that store, process, or transmit cardholder data, including merchants, service providers, and financial institutions.

- **Key requirements**: The standard consists of 12 requirements, organized into six control objectives, which outline various security measures that need to be implemented:

  - **Build and maintain a secure network**: This includes installing and maintaining firewalls, using secure configurations for network devices, and protecting cardholder data during transmission

  - **Protect cardholder data**: Encryption, masking, and other security measures must be implemented to protect cardholder data at rest and in transit

  - **Maintain a vulnerability management program**: Regularly update and patch systems, use up-to-date antivirus software, and conduct vulnerability scans and penetration testing

  - **Implement strong access control measures**: Restrict access to cardholder data on a need-to-know basis, assign unique IDs to users, and implement two-factor authentication

  - **Regularly monitor and test networks**: Monitor all access to network resources, track and monitor access to cardholder data, and conduct security testing, including penetration testing

  - **Maintain an information security policy**: Develop and maintain a comprehensive security policy addressing information security for all personnel

- **Compliance validation**: Compliance with the PCI DSS is validated through various means:

  - **A self-assessment questionnaire (SAQ)**: Merchants and service providers complete an SAQ annually to assess their compliance based on their specific environment and requirements

  - **External vulnerability scans**: Organizations must conduct quarterly external vulnerability scans by an **Approved Scanning Vendor (ASV)**

  - **Onsite assessments**: Some organizations may require an annual on-site assessment by a **Qualified Security Assessor (QSA)**

- **Levels of compliance**: The PCI DSS categorizes entities into different levels based on the number of transactions processed annually. The compliance requirements may vary based on the level, ranging from self-assessment to more extensive assessments and audits.

- **Implications for cloud-native**: When implementing cloud-native applications, organizations must ensure that the **CSP** is PCI DSS-compliant and that the shared responsibility model is understood and followed. Organizations must also implement additional security controls within their cloud environments to meet PCI DSS requirements.

- **Non-compliance consequences**: Failure to comply with the PCI DSS can lead to severe consequences, including financial penalties, increased transaction fees, loss of card processing privileges, reputational damage, and legal liabilities.

As a compliance engineer, understanding PCI DSS requirements, implementing appropriate controls, and conducting regular assessments and audits are crucial to maintain the security of cardholder data and meet compliance obligations. It is essential to stay updated with changes in the standard and emerging security threats to ensure ongoing compliance and protection against potential risks.

The HIPAA

The HIPAA is a comprehensive set of regulations in the US that governs the security and privacy of **protected health information (PHI)**. As per CDC regulations, the HIPAA applies to covered entities, such as healthcare providers, health plans, and healthcare clearinghouses, as well as their business associates who handle PHI on their behalf. (You can read more about the HIPAA by going to the link provided in the *Further reading* section.)

To provide a comprehensive overview, let's explore the key aspects of the HIPAA and its implications for healthcare applications:

- **PHI**: The HIPAA defines PHI as any individually identifiable health information held or transmitted by a covered entity or its business associate. This includes demographic information, medical records, test results, health insurance information, and other data that can be linked to an individual's health condition.

- **Privacy rule**: The HIPAA Privacy Rule establishes standards for protecting individuals' medical records and other PHI. It outlines patients' rights over their health information, including the right to access, request amendments, and receive an accounting of disclosures. Covered entities must have appropriate policies and procedures in place to safeguard PHI and obtain patient consent for certain uses and disclosures.

- **Security rule**: The HIPAA Security Rule sets forth the requirements for protecting **electronic PHI (ePHI)**. It mandates administrative, physical, and technical safeguards to ensure the confidentiality, integrity, and availability of ePHI. Covered entities must implement security measures such as access controls, encryption, audit controls, and workforce training to safeguard ePHI.

- **Breach notification regulation**: The rule concerning breach notifications under the HIPAA mandates that entities covered by the act must alert those impacted, the **Health and Human Services Department (HHS)**, and occasionally the press, if there's a breach involving unprotected PHI. This regulation provides details on the requirements for notification and the timeframe for doing so, depending on the count of individuals involved.

- **Business associate requirements**: The HIPAA extends its compliance requirements to business associates, including vendors, contractors, and other entities that handle PHI on behalf of covered entities. Business associates must enter a written agreement with the covered entity, known as a **Business Associate Agreement (BAA)**, that outlines their responsibilities and compliance obligations.

- **Implications for healthcare applications**: Healthcare applications, especially those that handle and transmit PHI, must adhere to HIPAA requirements. This includes implementing strong access controls, encrypting ePHI, conducting regular risk assessments, and ensuring secure data storage and transmission. Application developers and service providers must enter into BAAs with covered entities and follow the security and privacy provisions outlined by the HIPAA.

It is important to note that compliance with the HIPAA is not optional. Non-compliance can result in significant penalties, ranging from monetary fines to criminal charges, depending on the severity of the violation.

As a compliance engineer working with healthcare applications, it is crucial to have a deep understanding of HIPAA requirements, conduct regular risk assessments, and implement robust security measures to protect PHI. Staying informed about updates to the HIPAA regulations and industry best practices is vital to maintain compliance and ensure the privacy and security of patients' health information.

FISMA

The **Federal Information Security Management Act (FISMA)** is a US federal law that establishes a comprehensive framework for managing the security of federal information systems. Enacted in 2002, FISMA mandates specific security requirements, risk management practices, and reporting obligations for federal agencies.

To provide a comprehensive overview, let's explore the key aspects of FISMA and its implications for cloud-native software security in federal applications:

- **Security framework**: FISMA requires federal agencies to implement a risk-based approach to information security. It emphasizes the importance of developing and maintaining an agency-wide information security program that includes policies, procedures, and controls to protect federal information and systems.

- **Risk management**: FISMA mandates that federal agencies conduct periodic risk assessments to identify vulnerabilities, threats, and the potential impact of security breaches. These assessments help agencies make informed decisions about implementing appropriate security controls and mitigating risks.

- **Security controls**: FISMA requires federal agencies to implement a set of security controls, known as the **National Institute of Standards and Technology (NIST) Special Publication (SP)** 800-53, which provides a comprehensive catalog of controls covering various security domains. These controls address areas such as **access control**, **incident response**, **system and information integrity**, and **security awareness training**.

- **Continuous monitoring**: FISMA promotes the concept of continuous monitoring, whereby agencies continually assess, track, and remediate security vulnerabilities and incidents. This approach ensures that security controls are effective and provides timely visibility into potential risks.

- **Certification and accreditation**: FISMA requires federal agencies to establish a **certification and accreditation (C&A)** process to assess the security posture of their information systems. This process involves evaluating system security controls, documenting the results, and obtaining **authorization to operate (ATO)** before systems are deployed or undergo significant changes.

- **Reporting and compliance**: FISMA mandates that federal agencies report on their information security programs, including submitting annual reports to the **Office of Management and Budget (OMB)** and providing updates on security incidents and remediation efforts. Compliance with FISMA is monitored and enforced through audits and inspections conducted by the **Inspectors General (IG)** agency and other oversight bodies.

Compliance engineers need to have a deep understanding of FISMA and its associated NIST standards, such as **NIST SP 800-53**. They should stay updated on the evolving guidelines and recommendations from NIST to align their cloud-native security practices with FISMA requirements. Engaging in ongoing training and collaboration with agency stakeholders will support the effective implementation of FISMA and help ensure the security and resilience of federal information systems.

Case studies – incidents related to standards and their implications for security engineers

Let's investigate some of the most famous and devastating compliance and law violations in the past and understand the security lessons that we can take from them.

Target data breach and PCI DSS implications

In 2013, retail giant Target experienced a significant data breach that resulted in the compromise of approximately 40 million credit and debit card records. The incident highlighted the importance of compliance with the PCI DSS and its implications for security engineers.

The following are the implications for security engineers:

- **Security vulnerabilities:** The Target breach occurred due to a cybercriminal exploiting vulnerabilities in the company's network, gaining unauthorized access to sensitive cardholder data. Security engineers must proactively identify and address vulnerabilities to prevent such incidents.

- **PCI DSS compliance:** Target's breach led to scrutiny of its compliance with PCI DSS requirements. Security engineers play a critical role in implementing and maintaining the necessary security controls to meet PCI DSS standards, such as encryption, access controls, and regular security testing.

- **Incident response:** The incident highlighted the importance of a robust incident response plan. Security engineers should develop and test incident response procedures to minimize the impact of a breach and efficiently handle the aftermath.

Anthem data breach and HIPAA implications

In 2015, health insurance provider Anthem suffered a significant data breach, exposing the personal information of nearly 78.8 million individuals. The incident raised concerns about the security of PHI and compliance with the HIPAA.

The following are the implications for security engineers:

- **PHI protection:** The Anthem breach highlighted the critical need for security measures to protect PHI. Security engineers in healthcare applications must implement robust access controls, encryption, and other safeguards to ensure the confidentiality and integrity of PHI.

- **HIPAA compliance:** Anthem's breach triggered an investigation by the Department of Health and Human Services' **Office for Civil Rights (OCR)** to assess compliance with the HIPAA. Security engineers should work closely with compliance teams to ensure adherence to HIPAA requirements, such as conducting risk assessments, implementing security policies, and training employees on data security practices.

- **Breach response**: The incident emphasized the importance of a comprehensive breach response plan. Security engineers should collaborate with incident response teams to develop procedures for detecting, containing, and mitigating breaches promptly.

Office of Personnel Management (OPM) breach and FISMA implications

The **Office of Personnel Management (OPM)** data breach, discovered in 2015, exposed the sensitive personal information of millions of federal employees and applicants. The incident raised concerns about compliance with FISMA and the protection of federal systems and data.

The following are the implications for security engineers:

- **Risk management**: The OPM breach highlighted the importance of robust risk management practices. Security engineers should conduct thorough risk assessments, implement appropriate security controls, and regularly monitor and update systems to mitigate potential threats.

- **FISMA compliance**: The breach led to an assessment of OPM's compliance with FISMA requirements. Security engineers should ensure that federal applications meet FISMA's security control objectives, engage in continuous monitoring, and participate in certification and accreditation processes.

- **Supply chain security**: The incident underscored the significance of securing the supply chain. Security engineers should evaluate and validate the security practices of third-party vendors and contractors, particularly those handling sensitive government information.

These case studies illustrate the real-world implications of non-compliance or inadequate implementation of compliance standards. Security engineers must prioritize adherence to these standards, learn from past incidents, and proactively implement effective security controls and practices to protect data, systems, and individuals' privacy.

Navigating the legal and compliance landscape is essential in the cloud-native landscape. Understanding the key principles and implications of these regulations enables software engineers and security professionals to build secure and compliant cloud-native solutions. By staying updated, implementing robust security controls, and proactively addressing vulnerabilities, professionals can foster trust, protect data, and ensure compliance with legal requirements.

Summary

In this chapter, we covered a wide range of topics related to the legal and compliance aspects of cloud-native software security. We began by exploring privacy in the cloud, defining key terms, and examining the importance of privacy. We then delved into specific laws and regulations such as the CCPA, FTCA, CFAA, and HIPAA, discussing their key principles and real-world case studies. Next, we examined the significance of audit processes and methodologies in cloud-native adoption, emphasizing the importance of regular audits and showcasing a relevant case study. Furthermore, we provided a comprehensive overview of compliance standards, including SOC 2, PCI DSS, HIPAA, and FISMA, highlighting their implications for cloud-native software security. Real-world case studies demonstrated incidents related to these standards and their impact on security engineers.

By gaining knowledge in these areas, professionals can ensure compliance, strengthen security measures, and build trust with customers, stakeholders, and regulators. It is vital to stay updated with the evolving legal landscape and industry best practices to maintain the security and privacy of cloud-native applications.

Now that we understand the laws and doctrines pertaining to the cloud security space, in the next chapter we will further augment our understanding of cloud security by learning the vendor management and security certification and audit requirements that applications are expected to be conformant with.

Quiz

Answer the following questions to test your knowledge of this chapter:

- Scenario: A cloud-native software company handles the sensitive personal information of California residents. Which privacy law should they primarily focus on to ensure compliance?

- Imagine you are a security engineer responsible for auditing a cloud-native application. How would you approach identifying security risks and vulnerabilities specific to the cloud environment?

- Reflecting on the case study involving Target's data breach, what specific PCI DSS requirements could have been more effectively implemented to mitigate the risk of a cyberattack?

- Consider the implications of the Anthem data breach. As a security engineer working on healthcare applications, what additional security measures would you implement to ensure compliance with HIPAA and protect sensitive patient data?

- Putting yourself in the shoes of a compliance engineer working with federal applications, explain the steps you would take to ensure FISMA compliance and maintain the security of federal information systems.

Further readings

To learn more about the topics covered in this chapter, take a look at the following resources:

- `https://www.nist.gov/publications`
- `https://www.pcisecuritystandards.org/document_library`
- `https://www.hhs.gov/hipaa/for-professionals/guidance/index.html`
- `https://www.nist.gov/itl/fisma`
- `https://cloudsecurityalliance.org/research/`
- `https://gdpr-info.eu/art-4-gdpr/`
- `govinfo.gov/content/pkg/CHRG-115hhrg27917/pdf/CHRG-115hhrg27917.pdf.`
- `https://www.ftc.gov/business-guidance/privacy-security/gramm-leach-bliley-act`
- `https://cdn.ca9.uscourts.gov/datastore/opinions/2016/07/05/14-10037.pdf`
- `https://listings.pcisecuritystandards.org/documents/pci_ssc_quick_guide.pdf`
- `https://www.cdc.gov/phlp/publications/topic/hipaa.html`

10

Cloud Native Vendor Management and Security Certifications

Cloud computing has radically changed the way organizations manage their IT infrastructure, bringing with it unparalleled flexibility and scalability. Yet, the move to the cloud has also introduced new challenges related to **vendor management** and security. Managing vendor relationships effectively is critical to mitigating risks, ensuring compliance, and achieving your organization's broader enterprise risk management objectives. This chapter dives deep into the world of cloud vendor management and security certifications, revealing practical tools and strategies to build strong vendor relationships that underpin secure cloud operations.

In the previous chapter, we delved into the myriad of legal and compliance issues that organizations need to navigate in the world of cloud computing, from US state and federal laws to industry-specific regulations. Vendor management is a natural extension of this topic. The laws, regulations, and standards that apply to your organization also extend to your vendors. As such, you must ensure that they are not only aware of these obligations but are also fully compliant. By the end of this chapter, you will understand the various risks associated with cloud vendors and how to assess a vendor's security posture effectively. You will gain an appreciation for the role of security policy frameworks, government and industry cloud standards, and vendor security certifications in ensuring cloud security. Furthermore, you will be introduced to practical steps for risk analysis and best practices for vendor selection. We will also explore the art of building and maintaining healthy vendor relationships, emphasizing the importance of transparency and open communication. Finally, you will see how all of these concepts are applied in real-world scenarios through a comprehensive case study.

This knowledge will empower you to navigate your cloud vendor relationships effectively and align with the overall security strategy and risk management objectives of your organization. It's about turning potential risks into opportunities for enhanced security and compliance in your cloud operations.

In this chapter, we're going to cover the following main topics:

- Security policy framework
- Government cloud standards and vendor certifications
- Enterprise risk management
- Risk analysis
- Case study

Security policy framework

As organizations increasingly leverage the power of cloud computing, the complexity of managing relationships with cloud vendors also rises. These relationships require careful attention as they are an integral part of your organization's broader enterprise risk management program.

Cloud vendor management is about much more than just overseeing **service-level agreements (SLAs)**. It's about ensuring that your cloud vendors align with your business objectives, comply with relevant regulations, and provide the necessary security to protect your sensitive data and systems. This involves a comprehensive understanding of the vendor's services, their compliance with industry standards, their approach to risk management, and their overall capability to meet your organization's requirements.

Effective vendor management plays a pivotal role in an organization's enterprise risk management program. It's about identifying, assessing, and managing the risks associated with your cloud vendors. By having a robust vendor management process in place, you can ensure that your vendors don't become your weakest link in terms of security and compliance.

As we delve into this chapter, we'll explore the practical tools and techniques for managing vendor relationships effectively, assessing their security posture, and integrating them into your enterprise risk management program. This will not only help you minimize your risk exposure but also ensure that you fully leverage the benefits that cloud computing can bring to your organization. It is not uncommon for companies to lock in with a cloud vendor only to later realize that their security standards and policies are not at par with the government regulations that they need to be conformant to, so it becomes important to understand the security risks that arise when associating with a cloud vendor. Let's look at a few of them.

Understanding cloud vendor risks

In the cloud era, organizations increasingly rely on vendors to perform functions that were traditionally managed in-house. While these relationships can bring about significant benefits, they also introduce new risks. To manage these effectively, it is crucial to understand the types of risks associated with cloud vendors and the importance of assessing their security posture.

Types of risks associated with cloud vendors

When engaging with a cloud vendor, you are effectively extending your organization's boundary to include their services. This expanded boundary introduces several types of risks that must be managed carefully. They are as follows:

- **Data breaches**: One of the most serious risks associated with cloud vendors is the potential for data breaches. In a cloud environment, your data resides on shared infrastructure, which, if not properly secured, could be accessed by unauthorized parties. The impact of such breaches can be devastating, leading to financial losses, reputational damage, and legal repercussions.

- **Service availability**: The nature of cloud services means that they are expected to be available at all times. However, interruptions can occur due to technical failures, maintenance, security incidents, or other reasons. The risk of service downtime, even if it is minimal, can be a significant concern, particularly for critical applications or services.

- **Vendor lock-in**: Vendor lock-in refers to the difficulty in moving your services or data from one vendor to another. This can occur due to proprietary technology, contractual limitations, or the lack of interoperable standards. Vendor lock-in can limit your flexibility and potentially lead to higher costs or unsatisfactory service levels.

- **Compliance risks**: Cloud vendors may not always comply with the same regulatory standards as your organization, and this can expose your organization to compliance risks. For example, data privacy regulations often require that data is stored and processed in certain ways or in specific locations. If a vendor fails to meet these requirements, it could result in penalties for your organization.

- **Supply chain risks**: Cloud vendors often rely on their own suppliers, which introduces additional risks. If a supplier fails to deliver, it could impact the cloud vendor's service, and by extension, your organization.

Importance of assessing vendor security posture

Given the potential risks associated with cloud vendors, it is crucial to assess a vendor's **security posture** before engaging with them and regularly thereafter. Assessing a vendor's security posture involves evaluating their security policies, practices, and infrastructure to ensure they can protect your data and services adequately. Here are a few security tactics that you should look into when you're considering purchasing a vendor product and assessing their security posture:

- **Security policies and practices**: Review the vendor's security policies to ensure they align with your organization's standards and industry best practices. Look for robust policies covering areas such as access control, encryption, incident response, and disaster recovery. The vendor's employees should also undergo regular security training to ensure they understand and follow these policies.

- **Technical security measures**: Examine the technical security measures the vendor has in place. These should include firewalls, intrusion detection and prevention systems, data encryption, and regular vulnerability scanning and patching. It is also crucial to consider how the vendor isolates customer data, particularly in a multi-tenant environment.

- **Compliance certifications**: Compliance certifications can provide some assurance of a vendor's security posture. Look for certifications such as **ISO 27001** or **SOC 2**, which indicate the vendor has met internationally recognized security standards.

- **Incident response and disaster recovery**: The vendor should have robust incident response and disaster recovery plans. These should detail how they will respond to a security incident or disaster, how they will communicate with customers, and how they will restore services.

- **Third-party audits**: Consider requiring the vendor to undergo regular third-party audits. These can provide an unbiased assessment of the vendor's security posture and help identify any potential weaknesses.

Understanding and managing the risks associated with cloud vendors is a complex but crucial task. It requires a comprehensive understanding of the potential risks and a thorough assessment of the vendor's security posture. By doing so, organizations can reap the benefits of cloud services while minimizing their risk exposure.

Understanding security policy frameworks

As organizations move their operations into the cloud, the role of a **security policy framework** becomes even more critical. A well-defined and properly implemented security policy framework is vital to safeguard the organization's digital assets. It provides a strategic direction and clearly defined boundaries for all security-related activities within the organization. When working with cloud vendors, the policy should encompass not only the organization's internal environment but also the services provided by the cloud vendor.

The purpose and structure of a security policy framework

The primary purpose of a security policy framework is to establish a set of standards, guidelines, and practices that protect the confidentiality, integrity, and availability of an organization's information assets. This framework is usually a documented set of principles and rules that guide decision-making concerning the use of IT infrastructure, especially those related to cybersecurity.

In terms of structure, a security policy framework is typically composed of several layers:

- **Security policy**: This is a high-level document that outlines the overall security objectives of an organization, the responsibilities of different stakeholders, and the general approach to managing security risks.

- **Standards**: These are mandatory requirements that provide a detailed description of what must be done to comply with the security policy. For example, a standard might specify the type of encryption to be used for data transmission.

- **Guidelines**: These are non-mandatory recommendations that support the standards and policies. They provide advice on how to implement the standards effectively.

- **Procedures**: These are step-by-step instructions that describe how specific tasks should be performed. Procedures are often used to support compliance with the standards.

Implementing security policy frameworks with cloud vendors

Implementing a security policy framework with cloud vendors can be complex as you need to ensure that the vendor's practices align with your policies and standards. Here are some steps to consider:

- **Alignment with the vendor**: It's essential to ensure that your security policy framework aligns with your cloud vendor's practices. This requires a thorough understanding of the vendor's security policies, procedures, and controls.

- **Incorporate vendor contracts**: The requirements of your security policy framework should be incorporated into contracts with cloud vendors. This helps to establish clear expectations and responsibilities for both parties. It also provides a legal basis for enforcement if the vendor fails to meet the specified security standards.

- **Monitoring and auditing**: Once the policy is in place, it's important to regularly monitor the vendor's compliance with the policy. This can be achieved through **periodic audits**, **vulnerability assessments**, and **incident reports**.

Effective security policy framework in a cloud environment

Let's take a look at a case study of a global financial firm that successfully implemented a security policy framework with its cloud vendors. Given the highly regulated nature of the financial industry, the firm had to ensure that its vendors could meet stringent security requirements.

The firm began by developing a comprehensive security policy framework, which included policies, standards, guidelines, and procedures tailored to the specific risks of the cloud environment. This framework was developed in collaboration with IT, business stakeholders, and the legal team to ensure it addressed all relevant regulatory requirements and business needs.

Once the framework had been established, the firm engaged with its cloud vendors to align their practices with the firm's security policy. They incorporated security requirements into vendor contracts and established clear lines of responsibility for security tasks.

The firm also set up a robust monitoring system to ensure continuous compliance with the security policy framework. They used a combination of automated tools and periodic audits to identify any deviations from the policy and address them promptly.

In the end, the firm was able to successfully navigate the complexities of managing security in a multi-vendor cloud environment. Through its comprehensive security policy framework, it managed to mitigate risks, meet compliance requirements, and maintain a robust security posture.

It is also important to note that their security policy framework implementation was not an overnight task; it required careful planning, persistent execution, and continuous monitoring. The total execution included the following steps:

1. **The planning phase**: The firm spent considerable time in the planning phase. They conducted comprehensive risk assessments to understand the specific security threats related to their cloud environment. These assessments helped them design a security policy framework that was tailored to their unique risk profile.

2. **The execution phase**: During the execution phase, the firm worked closely with its cloud vendors. They made sure the vendors understood the security requirements and integrated these requirements into their practices. They also ensured the contract with each vendor reflected the responsibilities and expectations defined in the security policy framework.

3. **The monitoring phase**: In the monitoring phase, the firm established regular audits to ensure ongoing compliance with the security policy framework. They also set up an incident response plan to handle any security incidents effectively.

The key lesson from this case study is that implementing a security policy framework with cloud vendors is not a one-time activity. It requires ongoing efforts to ensure that the policy remains effective in the face of evolving security threats and business requirements.

Best practices for implementing a security policy framework with cloud vendors

While the steps discussed earlier provide a foundational approach, there are several best practices that organizations should consider when implementing their security policy framework with cloud vendors. They are as follows:

- **Define clear roles and responsibilities**: It's essential to establish clear roles and responsibilities between your organization and the cloud vendor. The security policy framework should define who is responsible for implementing and monitoring each aspect of the security controls.

- **Collaborate with the cloud vendor**: Successfully implementing a security policy framework requires a partnership approach. Engage with the cloud vendor during the development of the policy – their expertise in their specific technologies can be invaluable in setting realistic and effective security controls.

- **Maintain ongoing communication**: Regular communication is key to managing the relationship with the cloud vendor. Regular meetings can help address any security concerns promptly and ensure that the vendor remains aligned with your security policy framework.

- **Create a shared understanding of risk**: Both your organization and the cloud vendor should have a shared understanding of the potential security risks. This common understanding can help ensure that both parties take the necessary steps to mitigate risks.

- **Review and update the policy regularly**: The security policy framework should not be static. It should be reviewed and updated regularly to address changes in the threat landscape, regulatory requirements, or business operations.

Now that we understand the importance and the need for security certifications, let's learn about some of the standards that have been set out by the government and other vendor organization bodies that companies need to comply with to either gain the certifications or continue holding them.

Government cloud standards and vendor certifications

Navigating the world of cloud services is a complex task, one that is made even more intricate due to the various government and industry standards that organizations must adhere to. These standards, formulated by governmental bodies and industry groups, establish a baseline for security, interoperability, and data privacy in the cloud.

Governmental bodies worldwide have recognized the need for standards and regulations to govern the use of cloud services, especially considering the potential risks associated with data security and privacy. These standards aim to create an environment where cloud services can be utilized safely and effectively. Here are some significant government standards:

- **Federal Risk and Authorization Management Program (FedRAMP)**: This is a US government-wide program that standardizes the approach to security assessment, authorization, and continuous monitoring for cloud services. FedRAMP ensures that cloud services meet stringent security requirements before they are adopted by federal agencies.

- **Cloud Security Alliance – Security, Trust, Assurance, and Risk (CSA STAR)**: According to the Canadian Centre for Cyber Security, the CSA STAR program is a powerful *"program encompassing key principles of transparency, rigorous auditing, and harmonization of standards."* CSA STAR certification provides multiple benefits, including indicating to customers that a cloud service provider is using best practices and is transparent with security processes.

- **National Institute of Standards and Technology (NIST) guidelines**: NIST provides several guidelines related to cloud computing, including **NIST SP 800-145**, which defines cloud computing, **NIST SP 800-146**, which provides a cloud computing synopsis and recommendations, and **NIST SP 500-292**, which provides a cloud computing reference architecture and taxonomy.

Industry cloud standards

Industry groups have also created standards that provide guidelines for cloud service providers and users. These standards often focus on specific areas of cloud computing, such as security, interoperability, and portability. Here are some of the key industry standards:

- **ISO/IEC 27017**: This standard offers guidance on the facets of information security in the realm of cloud computing. It advocates for the establishment of cloud-specific information security controls that enhance the recommendations of the ISO/IEC 27002 and ISO/IEC 27001 standards, which focus on the management systems for information security.

- **ISO/IEC 27018**: This standard sets universally recognized control objectives, controls, and guidelines for putting into action measures to safeguard **Personally Identifiable Information (PII)** in line with the privacy principles outlined in ISO/IEC 29100, specifically within the context of the public cloud computing environment.

The importance of adhering to government and industry cloud standards

Adherence to these standards is not just about compliance; it's about ensuring the robustness and safety of your organization's cloud environment. By adhering to government and industry cloud standards, organizations can ensure that their cloud services are secure, reliable, and efficient.

Furthermore, adherence to these standards can foster trust between your organization and your customers, stakeholders, and regulatory bodies. It shows that you take the security and privacy of your cloud environment seriously and are committed to following best practices to protect it.

Let's expand upon these standards and delve deeper into what they entail and how they influence cloud vendor management.

Exploring the Federal Risk and Authorization Management Program (FedRAMP)

FedRAMP, introduced by the US government, has been designed to support the federal government's *Cloud-First* initiative by standardizing security assessment and accreditation. It's a risk management program that aims to ensure cloud products and services used in federal agencies meet rigorous security standards.

FedRAMP involves three main participants: a **cloud service provider** (**CSP**), an independent assessor known as a **Third-Party Assessment Organization** (**3PAO**), and the **program management office** (**PMO**), which oversees the process. The CSP undergoes a security assessment by the 3PAO, which verifies whether the provider meets the necessary security controls. The PMO, along with a **Joint Authorization Board**, reviews the assessment and can grant a provisional authority to operate. This allows federal agencies to leverage the authorization instead of conducting their own, separate assessments.

As such, for organizations serving federal agencies or dealing with government data, compliance with FedRAMP is not just a good-to-have; it's often a necessity. And for others, it's still a valuable benchmark indicating that a CSP is serious about security.

The role of the Cloud Security Alliance (CSA) STAR program

The CSA STAR program is a robust offering that covers all major areas of cloud computing security. It provides a comprehensive approach to assessing the security posture of cloud providers, integrating a **consensus-driven cloud-specific security control matrix** (**CSA CCM**) with a recognized third-party assessment and certification process.

The STAR program's three levels – **Self-Assessment**, **STAR Certification**, and **STAR Attestation/ Continuous** – provide progression for cloud providers to gradually improve their security posture. For organizations choosing cloud vendors, the CSA STAR certification can serve as a robust sign of a provider's commitment to transparency and best practices in cloud security.

Unpacking the National Institute of Standards and Technology (NIST) guidelines

NIST, a non-regulatory federal agency within the US Department of Commerce, is known for its technology standards, including a suite of standards and guidelines that are essential for cloud computing.

NIST's cloud computing standards provide definitions, taxonomies, and architecture that can be universally applied, helping to create a common language around cloud computing. NIST's cloud security guidelines further offer a detailed perspective on cloud security and risks, providing useful insights for organizations moving their operations to the cloud.

Understanding the ISO/IEC 27017 and ISO/IEC 27018 standards

The **International Organization for Standardization (ISO)** and the **International Electrotechnical Commission (IEC)** have also created critical standards for cloud security.

ISO/IEC 27017 provides a code of practice for information security controls for cloud services. It's an extension to **ISO/IEC 27002**, tailored for cloud services. This standard can help organizations understand what they should be looking for in terms of information security when choosing a CSP.

ISO/IEC 27018 focuses on the protection of personal data in the cloud. It establishes a set of guidelines and principles for ensuring PII processed by a public CSP is protected according to recognized privacy principles.

By understanding these standards in depth and adhering to them, organizations can ensure they're not just ticking compliance boxes but are also truly enhancing their cloud security posture. These standards serve as the golden thread that ties together the entire fabric of an organization's cloud security strategy, providing the necessary guidance, direction, and continuity.

Vendor certifications

As organizations increasingly move their operations to the cloud, the importance of vendor security certifications has grown significantly. These certifications play a crucial role in assessing the security risks associated with a given CSP. They offer proof that a vendor has met a certain standard of security practices and controls, as determined by an unbiased, third-party auditor. Let's do a deep dive into the security certifications that you should be aware of that exist and hold gravitas in the merit of holding those certifications.

The role of vendor certifications in assessing security risk

Cloud vendor certifications are essential in assessing the security posture of potential vendors. They offer a level of assurance that the vendor has been thoroughly evaluated against a defined set of criteria and found to meet the standards for data protection and privacy.

By evaluating the certifications held by potential cloud vendors, an organization can gain a better understanding of the vendor's commitment to security, the maturity of its security controls, and its ability to protect sensitive data. Certifications provide a benchmark for comparison, allowing organizations to weigh up the security capabilities of different vendors.

Moreover, certifications can also be beneficial in meeting regulatory requirements. Many industries are subject to specific regulations around data security, and using certified vendors can help organizations comply with these requirements.

Common security certifications to look for in cloud vendors

There are several common security certifications that organizations should consider when evaluating potential cloud vendors. They are as follows:

- **System and Organization Controls (SOC) 2**: SOC 2 is a certification designed for technology and cloud computing companies. It is based on the **AICPA's Trust Service Criteria** and assesses the extent to which a vendor complies with one or more of the following principles: security, availability, processing integrity, confidentiality, and privacy.

- **ISO 27001**: This is an internationally acknowledged benchmark for the management of information security. It outlines the prerequisites for the creation, execution, upkeep, and constant enhancement of an information security management system. Providers that have achieved the ISO 27001 certification have shown a methodical strategy for handling confidential data and guaranteeing the security of such information.

- **FedRAMP**: As mentioned in the previous section, this is a US government-wide program that provides a standardized approach to security assessment, authorization, and continuous monitoring for cloud products and services. For organizations serving federal agencies, a FedRAMP certification is often essential.

Understanding what certifications cover and their limitations

While certifications provide an excellent starting point in evaluating a cloud vendor's security, it's crucial to understand what these certifications cover and their limitations.

Security certifications typically cover a defined set of controls and practices. For instance, a SOC 2 certification evaluates a vendor based on the **Trust Service Criteria** relevant to the services they offer. An ISO 27001 certification verifies that the vendor has implemented an information security management system in line with the standard's requirements.

However, certifications do not guarantee that a vendor is immune to all security risks. They are a snapshot in time, showing that the vendor met certain criteria at the point of assessment. They do not provide a guarantee of future security performance, nor do they cover all possible security risks.

Furthermore, not all certifications are created equal. Some have more rigorous requirements than others, and the value of a certification depends on the robustness of the underlying standard and the rigor of the assessment process.

Therefore, while certifications are a critical tool in assessing a cloud vendor's security, they should be only one aspect of a broader vendor assessment process. This process should also consider other factors, such as the vendor's security architecture, data privacy practices, and the specifics of what the vendor is offering in terms of SLAs and contractual obligations.

Vendor security certifications are a crucial piece of the puzzle in managing cloud security risks. They provide valuable insights into a vendor's security practices and can play a significant role in decision-making processes. However, it's essential to delve beyond the certifications and understand their scope, limitations, and what they mean in the broader context of your organization's security needs.

Vendor certifications as a litmus test for security preparedness

Vendor certifications can act as a litmus test for the overall preparedness of a cloud vendor in addressing security concerns. Cloud vendors that invest time and resources to obtain reputable certifications often have a strong culture of security and are more likely to have robust systems and processes in place to deal with threats. Hence, the presence of recognized certifications can be an initial indication of the vendor's level of commitment to data protection and regulatory compliance.

Interpreting certifications relative to business needs

A vendor may have numerous certifications, but they might not all be relevant to your business needs or compliance requirements. For instance, a healthcare organization would want to work with a cloud vendor with HIPAA certification, which is specifically designed for entities handling protected health information. Similarly, a business handling payment card data would look for a vendor with PCI DSS certification.

It is essential to understand which certifications are relevant to your organization and prioritize those when assessing vendors.

Vendor self-assessment questionnaires (SAQs)

Aside from certifications, another tool that can be useful in vendor assessment is the vendor **self-assessment questionnaire (SAQ)**. This is a set of questions provided to vendors, asking them to detail their security, privacy, and compliance controls. The **Cloud Security Alliance (CSA)**, for instance, offers a **Consensus Assessments Initiative Questionnaire (CAIQ)** that organizations can use to identify areas of concern and evaluate the security posture of their cloud vendors in a standardized way.

Continuous monitoring and review

Finally, it is crucial to note that vendor assessment is not a one-time process. Just as your organization's security landscape evolves, so too does that of your cloud vendors. Continually monitoring your vendors, staying updated on their certification status, and regularly reviewing their security and compliance controls are essential practices. This also involves staying informed about any significant changes to the certification standards themselves.

By leveraging certifications effectively and understanding their role in the broader context of security risk management, organizations can make informed decisions about their cloud vendors and contribute significantly to their overall cloud security strategy.

While the majority of what we have covered works perfectly fine at the org level or for the security/compliance team within an organization, let's understand other aspects that come into play when dealing with compliance at the enterprise level.

Enterprise risk management

Enterprise risk management (ERM) is a strategic discipline that involves identifying potential risks that may affect an organization and implementing appropriate mitigation strategies to reduce these risks. It's a holistic approach to risk management that considers a wide array of risks across the organization, including operational, financial, reputational, strategic, and, notably for this discussion, technological risks.

In today's digital age, where most businesses have a significant portion of their operations reliant on technology, specifically cloud-based services, ERM plays a crucial role in ensuring that potential security risks are adequately managed. Cloud environments introduce a unique set of risks, such as data breaches, system outages, vendor lock-ins, or compliance issues, that can have significant implications for business continuity and reputation.

Effective ERM in the cloud involves conducting a comprehensive risk assessment to identify and understand potential vulnerabilities and threats associated with the use of cloud services, followed by the development of a risk management plan to address these identified risks. This plan should include strategies for risk avoidance, mitigation, transfer, or acceptance, depending on the nature and severity of the risk.

The primary goal of ERM in cloud security is to ensure that risks are managed in a way that aligns with the organization's risk appetite and business objectives. It's about making informed decisions on which risks to take, which ones to avoid, and how best to manage those that are inevitable.

The significance of ERM in cloud security

The importance of ERM in cloud security cannot be overstated. Firstly, ERM provides a framework for identifying and understanding the risks associated with cloud services. This includes not only technical risks such as data breaches or system outages but also strategic risks such as vendor lock-in and reputational risks resulting from potential data privacy issues.

Secondly, ERM promotes a proactive approach to managing these risks. By considering risks in advance, organizations can put measures in place to prevent or mitigate potential threats, rather than simply reacting to them after the fact.

Thirdly, ERM supports compliance with regulatory requirements. Many industries are subject to specific regulations regarding data security and privacy, and an effective ERM program can help ensure that these requirements are met.

Finally, and perhaps most importantly, ERM helps organizations achieve their business objectives in a risk-controlled environment. By effectively managing cloud security risks, organizations can harness the power of the cloud to drive innovation and business growth, without compromising security or compliance.

Incorporating vendor management into your enterprise risk management program

While effective ERM in the cloud begins with understanding the risks associated with cloud services, it doesn't end there. A critical part of ERM is managing relationships with cloud vendors as they play a significant role in an organization's risk profile.

Cloud vendor management involves evaluating and selecting vendors based on their ability to meet your organization's security and business requirements, negotiating contracts that align with your risk appetite, and monitoring vendors' performance and compliance over time.

Incorporating vendor management into your ERM program involves several key steps, which are as follows:

1. **Vendor selection**: This begins with understanding your organization's risk tolerance and business requirements and using these as criteria for evaluating potential vendors. It involves assessing vendors' security capabilities, the robustness of their control environments, and their commitment to security, as demonstrated by relevant certifications.

2. **Contract negotiation**: Contracts with cloud vendors should be negotiated with your organization's risk profile in mind. This means ensuring that contracts include clear terms regarding data ownership, security responsibilities, right to audit, incident response, and liability in the event of a breach.

3. **Continuous monitoring**: The vendor management process does not end once a vendor is selected. Continuous monitoring is crucial to ensure that vendors are meeting their contractual obligations and maintaining a robust security posture. This may involve regular security audits, reviewing security incident reports, and monitoring **key performance indicators (KPIs)**.

4. **Vendor risk assessment**: Regular risk assessments should be conducted to identify any changes in the risk profile associated with a particular vendor. This should include an evaluation of the vendor's security controls and incident response capabilities, as well as compliance with relevant standards and regulations.

Incorporating vendor management into your ERM program is not just about minimizing risk, but also about maximizing value. By effectively managing cloud vendors, you can ensure that you're deriving the maximum benefit from your cloud services while maintaining a robust and compliant security posture.

Risk analysis

Risk analysis and vendor selection are fundamental aspects of effective cloud vendor management. The process of choosing a cloud vendor is no longer merely a technical decision but involves a thorough evaluation of the vendor's risk profile, security posture, and overall ability to align with your organization's business goals and risk appetite.

Risk analysis – a key step in vendor evaluation

Risk analysis is the process of identifying, assessing, and prioritizing risks, providing a basis for decision-making. In the context of cloud vendor selection, it helps organizations understand the risks they may inherit from their cloud vendors, enabling them to make informed decisions.

The steps involved in performing risk analysis on potential cloud vendors include the following:

1. **Identify potential vendors**: Compile a list of potential vendors based on your organization's cloud needs. This list should be broad enough to ensure a wide selection but should also include vendors that specialize in the kind of cloud services your organization needs.

2. **Assess vendor capabilities**: Review each vendor's capabilities in light of your organization's needs. Consider factors such as the robustness of their infrastructure, their service offerings, their scalability, and their ability to meet your specific business requirements.

3. **Identify risks associated with each vendor**: Identify the specific risks associated with each vendor. These could range from security risks such as data breaches or service availability issues to strategic risks such as vendor lock-in.

4. **Assess the severity and likelihood of each risk**: For each identified risk, assess its potential impact on your organization and its likelihood of occurrence. This will help prioritize the risks and focus on those that are most significant.

5. **Identify risk mitigation measures**: For each major risk, identify potential risk mitigation measures. These could range from technical controls to contractual measures.

6. **Evaluate the residual risk**: After considering the potential mitigation measures, evaluate the residual risk associated with each vendor – that is, the risk that remains after all mitigation measures are considered.

Tools and techniques for evaluating vendor risk

Several tools and techniques can assist in evaluating vendor risk:

- **Vendor SAQs**: SAQs can provide valuable insights into a vendor's security posture. They require the vendor to provide information about their security controls and practices. However, they rely on the vendor's honesty and understanding of their security environment.

- **Third-party audits and certifications**: These can provide a more objective assessment of a vendor's security posture. Look for vendors who hold recognized certifications such as ISO 27001 and SOC 2, or those relevant to your industry, such as **HIPAA** for healthcare or **PCI DSS** for organizations handling credit card data.

- **Security scorecard services**: Several firms provide security scorecard services that offer a quick assessment of a vendor's security posture based on publicly available information.

- **Penetration testing and security audits**: With the vendor's permission, you might also consider engaging a third party to conduct a security audit or penetration testing.

Best practices for vendor selection

The following are some best practices for vendor selection:

- **Define your business and security requirements**: Have a clear understanding of your business and security requirements. This will serve as a basis for evaluating potential vendors.

- **Look beyond price**: While cost is an important consideration, it shouldn't be the only factor. Other factors such as the vendor's security posture, their track record, their ability to meet your specific business needs, and their commitment to customer service are equally, if not more, important.

- **Evaluate the vendor's track record**: Look for a vendor with a strong track record in delivering secure, reliable services.

- **Seek references**: Ask potential vendors for references from customers in similar industries and with similar use cases.

- **Ensure a clear contract**: Make sure all security requirements, responsibilities, and obligations are clearly stated in the contract.

Building and managing vendor relationships

Once a cloud vendor has been selected, the focus shifts to building and managing the vendor relationship. Establishing a robust and mutually beneficial relationship with your cloud vendor can be the key to successful cloud deployments and operations.

Techniques for establishing strong vendor relationships

Establishing a strong relationship with your cloud vendor involves more than just signing a contract. It requires an ongoing commitment to communication, collaboration, and mutual understanding. Here are a few techniques:

- **Collaborative approach**: Instead of viewing the vendor as merely a service provider, treat them as a partner. This promotes a collaborative approach where both parties are invested in each other's success.

- **Open communication**: Regular and open communication helps foster mutual understanding and trust. This should include not just formal reporting but also regular informal interactions.

- **Shared goals and objectives**: Ensure that your vendor understands your business goals and objectives and how their services support these. This promotes alignment and ensures that both parties are working toward the same goals.

- **Ongoing education**: As your business evolves, so too will your cloud needs. Providing ongoing education about changes in your business can help the vendor better align their services to your evolving needs.

The importance of transparency and communication

Transparency and communication are the bedrock of a successful vendor relationship. Transparency promotes trust, while effective communication ensures alignment.

Transparency involves sharing relevant information openly with the vendor. This includes not only technical information but also strategic information such as changes to business objectives or risk appetite. It also means being honest about concerns or issues as they arise. Effective communication, on the other hand, ensures that both parties are on the same page. This involves clearly communicating your needs and expectations and listening to the vendor's feedback and suggestions.

Ongoing vendor management

Building a strong vendor relationship is not a one-time activity but an ongoing process that requires continuous effort and attention. Key aspects of ongoing vendor management include periodic reviews, performance metrics, and contract renegotiations:

- **Periodic reviews**: Regular reviews of the vendor's performance can help identify any issues or areas for improvement. These reviews should consider both qualitative and quantitative measures and should involve input from all stakeholders.

- **Performance metrics**: Clearly defined and agreed-upon performance metrics can provide an objective basis for evaluating the vendor's performance. These could include measures such as system uptime, response times, or the number of security incidents.

- **Contract renegotiations**: Over time, your business needs may change, and the original contract may no longer be adequate. Regular contract renegotiations can ensure that the contract continues to align with your needs and expectations.

By taking a structured, proactive approach, organizations can ensure that their vendor relationships are strong, mutually beneficial, and aligned with their business objectives. This not only reduces the risk associated with cloud vendors but also maximizes the value derived from cloud services. Remember, a strong and collaborative vendor relationship can greatly enhance the security, efficiency, and overall success of your cloud initiatives.

Now that we have gained a fair understanding of maintaining compliance both at an org level and enterprise level, let's look into a hypothetical case study so that we can put to use what we have just learned.

Case study

To illustrate the principles of cloud vendor management in a real-world context, let's consider the case of XYZ Corp, a hypothetical global e-commerce company, and their successful relationship with their primary cloud vendor, CloudTech (a hypothetical cloud vendor).

Background

XYZ Corp, as an e-commerce entity, processes millions of transactions daily. They needed a robust cloud environment to handle this volume while ensuring data security and system availability. In 2022, they chose CloudTech, one of the leading CSP globally, known for its robust infrastructure, scalability, and advanced security controls.

Risk analysis and vendor selection

XYZ Corp had a structured vendor selection process in place. They evaluated several vendors on different parameters, including service offerings, scalability, reliability, cost, and, most importantly, their security posture. They performed a thorough risk assessment, which included reviewing CloudTech's security certifications (SOC 2 and ISO 27001) and third-party audit reports. They also engaged an independent firm to perform a security audit. After considering the overall risk and benefits, they chose CloudTech.

Establishing strong vendor relationship

From the onset, XYZ Corp treated CloudTech not merely as a vendor but as a strategic partner. They ensured that CloudTech understood their business goals and risk appetite. There was open communication with regular meetings between the teams. The relationship was marked by transparency, collaboration, and mutual trust.

Managing the relationship

XYZ Corp's approach to vendor management was proactive and dynamic. They defined clear performance metrics that were tied to their business objectives. These included system availability, latency, and the number of security incidents. They conducted quarterly reviews of CloudTech's performance against these metrics. They also reviewed security incident reports and audit logs to monitor CloudTech's security posture.

Over the years, there were occasions when they had to renegotiate the contract to better align with their evolving needs. This was done collaboratively, ensuring a win-win outcome.

Successful outcomes

This proactive approach to vendor management enabled XYZ Corp to derive maximum benefit from their relationship with CloudTech. They were able to scale their operations significantly while maintaining a robust security posture. Over the years, they experienced no major security incidents and consistently met their system availability targets. The partnership with CloudTech became a key enabler of their business strategy, contributing to their global success.

This case study underlines the central thesis of this chapter – that effective cloud vendor management is a critical component of an organization's cloud security strategy. By carefully managing and continuously monitoring their relationships with cloud vendors, organizations can significantly mitigate risks and derive maximum value from their cloud initiatives.

Summary

In this chapter, we dove into the intricate world of cloud vendor management. We discussed the importance of understanding cloud vendor risks, such as data breaches and service availability, emphasizing the necessity of a thorough security posture assessment for potential vendors.

This chapter covered the structure and purpose of a security policy framework and its role in cloud environments. We explored government and industry cloud standards such as FedRAMP, GDPR, and ISO 27001 and their influence on cloud vendor management. The significance of vendor security certifications such as SOC 2 and ISO 27001 was underlined, offering you an understanding of their scope and limitations.

This chapter also detailed the integration of vendor management into enterprise risk management programs, underlining its impact on cloud security. Practical steps for risk analysis and vendor selection, as well as techniques for establishing and managing vendor relationships, were discussed.

Lastly, we illustrated these principles with a real-life case study of Capital One and AWS, showcasing how successful cloud vendor management can mitigate risks and bolster security. Overall, this chapter stressed that effective cloud vendor management is a continuous, strategic process integral to a robust cloud strategy.

Quiz

Answer the following questions to test your knowledge of this chapter:

- How can organizations strike a balance between leveraging the innovation and scalability of cloud vendors while mitigating the inherent risks associated with outsourcing critical business functions?

- In a rapidly evolving technology landscape, how can organizations ensure that their cloud vendor management practices remain adaptable to emerging risks and regulatory requirements?

- What steps can organizations take to foster a culture of transparency and open communication with their cloud vendors, promoting a collaborative approach that enhances security and trust?

- Considering the limitations of vendor security certifications, how can organizations effectively supplement their evaluation process to gain a comprehensive understanding of a vendor's security posture and commitment to data protection?

- Reflecting on the case study, what key lessons can be drawn from their successful vendor relationship in terms of risk assessment, contract negotiation, and ongoing management? How can these lessons be applied to other organizations to enhance their cloud vendor management practices?

Further readings

To learn more about the topics that were covered in this chapter, take a look at the following resources:

- https://cloudsecurityalliance.org/
- https://www.nist.gov/topics/cloud-computing
- https://aws.amazon.com/security/
- https://azure.microsoft.com/en-us/trust-center/
- https://cloud.google.com/security
- https://www.opengroup.org/cloud/cloud-computing
- https://www.iso.org/standard/43757.html
- https://cloudsecurityalliance.org/artifacts/security-guidance-v4/
- https://searchcloudsecurity.techtarget.com/definition/cloud-vendor-management
- https://www.fedramp.gov/

Index

A

www.packtpub.com

Subscribe to our online digital library for full access to over 7,000 books and videos, as well as industry leading tools to help you plan your personal development and advance your career. For more information, please visit our website.

Why subscribe?

- Spend less time learning and more time coding with practical eBooks and Videos from over 4,000 industry professionals

- Improve your learning with Skill Plans built especially for you

- Get a free eBook or video every month

- Fully searchable for easy access to vital information

- Copy and paste, print, and bookmark content

Did you know that Packt offers eBook versions of every book published, with PDF and ePub files available? You can upgrade to the eBook version at packtpub.com and as a print book customer, you are entitled to a discount on the eBook copy. Get in touch with us at customercare@packtpub.com for more details.

At www.packtpub.com, you can also read a collection of free technical articles, sign up for a range of free newsletters, and receive exclusive discounts and offers on Packt books and eBooks.

Other Books You May Enjoy

If you enjoyed this book, you may be interested in these other books by Packt:

DevSecOps in Practice with Vmware Tanzu

Parth Pandit, Robert Hardt

ISBN: 978-1-80324-134-0

- Build apps to run as containers using predefined templates
- Generate secure container images from application source code
- Build secure open source backend services container images
- Deploy and manage a Kubernetes-based private container registry
- Manage a multi-cloud deployable Kubernetes platform
- Define a secure path to production for Kubernetes-based applications
- Streamline multi-cloud Kubernetes operations and observability
- Connect containerized apps securely using service mesh

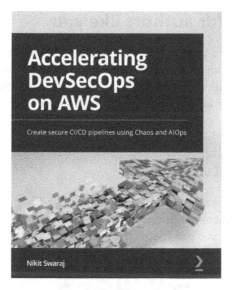

Accelerating DevSecOps on AWS

Nikit Swaraj

ISBN: 978-1-80324-860-8

- Use AWS Codestar to design and implement a full branching strategy
- Enforce Policy as Code using CloudFormation Guard and HashiCorp Sentinel
- Master app and infrastructure deployment at scale using AWS Proton and review app code using CodeGuru
- Deploy and manage production-grade clusters using AWS EKS, App Mesh, and X-Ray
- Harness AWS Fault Injection Simulator to test the resiliency of your app
- Wield the full arsenal of AWS Security Hub and Systems Manager for infrastructure security automation
- Enhance CI/CD pipelines with the AI-powered DevOps Guru service

Packt is searching for authors like you

If you're interested in becoming an author for Packt, please visit `authors.packtpub.com` and apply today. We have worked with thousands of developers and tech professionals, just like you, to help them share their insight with the global tech community. You can make a general application, apply for a specific hot topic that we are recruiting an author for, or submit your own idea.

Share Your Thoughts

Now you've finished *Cloud Native Software Security Handbook*, we'd love to hear your thoughts! Scan the QR code below to go straight to the Amazon review page for this book and share your feedback or leave a review on the site that you purchased it from.

`https://packt.link/r/1837636982`

Your review is important to us and the tech community and will help us make sure we're delivering excellent quality content.

Download a free PDF copy of this book

Thanks for purchasing this book!

Do you like to read on the go but are unable to carry your print books everywhere?

Is your eBook purchase not compatible with the device of your choice?

Don't worry, now with every Packt book you get a DRM-free PDF version of that book at no cost.

Read anywhere, any place, on any device. Search, copy, and paste code from your favorite technical books directly into your application.

The perks don't stop there, you can get exclusive access to discounts, newsletters, and great free content in your inbox daily

Follow these simple steps to get the benefits:

1. Scan the QR code or visit the link below

https://packt.link/free-ebook/9781837636983

2. Submit your proof of purchase
3. That's it! We'll send your free PDF and other benefits to your email directly

www.ingramcontent.com/pod-product-compliance
Lightning Source LLC
Chambersburg PA
CBHW062050050326
40690CB00016B/3039